Wear in Advanced Engineering Applications and Materials

Computational and Experimental Methods in Structures

ISSN: 2044-9283

Series Editor: Ferri M. H. Aliabadi *(Imperial College London, UK)*

This series will include books on state-of-the-art developments in computational and experimental methods in structures, and as such it will comprise several volumes covering the latest developments. Each volume will consist of single-authored work or several chapters written by the leading researchers in the field. The aim will be to provide the fundamental concepts of experimental and computational methods as well as their relevance to real world problems.

The scope of the series covers the entire spectrum of structures in engineering. As such it will cover both classical topics in mechanics, as well as emerging scientific and engineering disciplines, such as: smart structures, nanoscience and nanotechnology; NEMS and MEMS; micro- and nano-device modelling; functional and smart material systems.

Published:

More information on this series can also be found at http://www.worldscientific.com/series/cems

(Continued at end of book)

Computational and Experimental Methods in Structures – Vol. 12

Wear in Advanced Engineering Applications and Materials

Editors

Luis Rodríguez-Tembleque
Universidad de Sevilla, Spain

Jesús Vázquez
Universidad de Sevilla, Spain

M. H. Ferri Aliabadi
Imperial College London, UK

W🌐 World Scientific

NEW JERSEY · LONDON · SINGAPORE · BEIJING · SHANGHAI · HONG KONG · TAIPEI · CHENNAI · TOKYO

Published by

World Scientific Publishing Europe Ltd.

57 Shelton Street, Covent Garden, London WC2H 9HE

Head office: 5 Toh Tuck Link, Singapore 596224

USA office: 27 Warren Street, Suite 401-402, Hackensack, NJ 07601

Library of Congress Cataloging-in-Publication Data
Names: Rodriguez-Tembleque, Luis editor. | Vázquez, Jesús, editor. | Aliabadi, M. H., editor.
Title: Wear in advanced engineering applications and materials / editors, Luis Rodríguez-Tembleque,
    Universidad de Sevilla, Spain, Jesús Vázquez, Universidad de Sevilla, Spain,
    M. H. Ferri Aliabadi, Imperial College London, UK.
Description: New Jersey : World Scientific, [2022] | Series: Computational and experimental
    methods in structures, 2044-9283 ; vol. 12 | Includes bibliographical references and index.
Identifiers: LCCN 2021027315 (print) | LCCN 2021027316 (ebook) |
    ISBN 9781800610682 (hardcover) | ISBN 9781800610699 (ebook for institutions) |
    ISBN 9781800610705 (ebook for individuals)
Subjects: LCSH: Mechanical wear.
Classification: LCC TA418.45 .W43 2022 (print) | LCC TA418.45 (ebook) |
    DDC 620.1/1292--dc23
LC record available at https://lccn.loc.gov/2021027315
LC ebook record available at https://lccn.loc.gov/2021027316

British Library Cataloguing-in-Publication Data
A catalogue record for this book is available from the British Library.

For any available supplementary material, please visit
https://www.worldscienti ic.com/worldscibooks/10.1142/Q0315#t=suppl

Desk Editors: Christina Ramalingam/Michael Beale/Shi Ying Koe

Typeset by Stallion Press
Email: enquiries@stallionpress.com

Preface

Wear is one of the main reasons for inoperability in components having mechanical joints, and produces enormous costs to our society. An adequate understanding of the wear damage phenomenon allows engineers to accurately predict the operational life, or maintenance period, of modern mechanical elements, and thus reduce costs of inoperability or obtain an optimum design (i.e. selecting proper contact pair materials, shapes, and surface finishing according to the mechanical and environmental conditions, and desired durability). Therefore, the economic implication of a proper wear prediction in mechanical components can be of enormous value to the industry. This book presents a compilation of the main recent computation techniques and practical studies in the field of wear which are applied in advanced engineering applications and materials. Contributions from leading experts in these fields and younger promising researchers give a detailed overview of different modern techniques and materials that can be considered by researchers and practicing engineers in order to reduce surface damage and wear in mechanical components and materials. Also, this book is very useful for PhD students in Mechanical Engineering areas, especially to those research topics that are clearly governed by some type of surface damage due to the surface interactions produced by a mechanical contact: wear, fretting wear, fretting, and fretting fatigue. Finally, it serves also as a reference into the state of the art in the most recent numerical or experimental studies related to wear in practicing engineering applications and/or materials.

The editors are very grateful to all the contributors and all the staff at World Scientific Publishing for their encouragement, help, and support of producing this book, under the special circumstances caused by the coronavirus pandemic (COVID-19).

Luis Rodríguez-Tembleque
Jesús Vázquez
M. H. Ferri Aliabadi

About the Editors

Luis Rodríguez-Tembleque is an Associate Professor in the Department Continuum Mechanics and Theory of Structures at the Escuela Técnica Superior de Ingeniería (ETSI), Universidad de Sevilla, Spain. His research interests are focused on Computational Mechanics (based on the Finite Element Method and the Boundary Element Method), with emphasis on Computational Contact Mechanics (including friction, rolling-contact and wear simulation), Multiscale Material Modelling and Fracture and Damage Mechanics problems in advanced Multifield Materials (i.e. piezoelectric and magnetoelectroelastic materials) or Multifield Nanocomposites (i.e. carbon nanotube reinforced composites). He has co-edited several books on Wear and Contact Mechanics and he has published numerous chapters and papers in scientific journals and conference proceedings.

Jesús Vázquez is an Associate Professor in the Department of Mechanical Engineering and Manufacturing at the Escuela Técnica Superior de Ingeniería (ETSI), Universidad de Sevilla, Spain. He has worked mainly in fretting fatigue, fatigue in notched components and contact mechanics. In fatigue — fretting fatigue and fatigue at notches — he has worked mainly in the development of fatigue life models in order to predict the crack initiation and crack propagation phases. In fretting fatigue, he also has worked analysing the effect produced by shot and laser peening on the fretting fatigue life, being focused in the induced residual stress, its cyclic relaxation, and the surface roughness of these treatments. In contact mechanics, his research is focused on the development of analytical solutions for the contact stress/strain fields produced in different fretting cases.

M. H. Ferri Aliabadi is a Professor of Aerospace Structures and holds the Zaharoff Chair in Aviation at Imperial College London, UK. He was Head of Department of Aeronautics, Imperial College (2009–2017). Prior to joining Imperial College in 2005, he was Professor of Computational Mechanics and the Director of Aerospace Engineering at Queen Mary, University of London (1997–2004) and Reader and Head of Damage Tolerance Division at WIT, Southampton (1987–1997). Ferri M. H. Aliabadi has worked for 25 years in the field of Computational Structural Mechanics and has established an international reputation for his achievements in development of Computational Method related to Fracture & Damage Mechanics. He has pioneered a new generation of boundary element methods and is noted for his contributions to other fields including Acoustics, Nonlinear Mechanics, Wear Mechanics, Optimisation and Reliability Analysis and Multiscale Material Modelling.

Contents

Contents

Chapter 1

Fretting Fatigue and Wear of Spline Couplings: From Laboratory Testing to Industrial Application through Computational Modelling

Seán B. Leen

Mechanical Engineering, School of Engineering,
NUI Galway, H91 HX31, Ireland;
Ryan Institute for Environment, Marine and Energy, NUI Galway,
I-Form Centre for Advanced Manufacturing, Ireland
sean.leen@nuigalway.ie

One of the major challenges for fretting fatigue design is the bridging of the gap between laboratory tests, including associated theoretical and computer models, typically constrained to simplified loading conditions, and the need for real-life solutions, which relate more directly to in-situ loading, environmental and other relevant conditions, including geometrical constraints of target machine or structural components. This chapter describes experiences and challenges relating to the application of computational modelling to the design of simple but representative tests for fretting fatigue and wear of complex aeroengine spline couplings. It is proposed that the key step forward for fretting is to recognise, on the one hand, the importance of spatial and temporal field distributions of key multiaxial surface and sub-surface parameters, rather than try to reduce down to a single parameter or set of simple variables, such as normal load, coefficient of friction and stroke, and, on the other hand, to develop experimental tests which represent these spatial-temporal distributions. This is entirely feasible using modern nonlinear computational techniques, in combination with experimental and theoretical advances in fretting.

1. Introduction

The ongoing digitalisation of society presents enormous challenges, as well as opportunities, for engineering design to contribute to the global and local transformation of civilisation. Specific examples of global challenges include:

- Clean, efficient, sustainable energy.
- Abundant, affordable healthcare.
- Sustainable development and climate change.

Within this broader context, there are significant, specific scientific-technological challenges, more directly relevant to the topic of fretting. These include the development of *in-silico* design methods, viz. exploiting computer modelling for design; multi-scale modelling for improved design and safety; translating new micro-scale knowledge of materials and manufacturing into improved design of applications. For example, how can we translate the significant advances being made in high-magnification, high-resolution microscopy into improved design and manufacture of real-world components, devices, equipment and systems for improved, more sustainable, more cost-effective transport, living, healthcare, agriculture, industry, etc.? From a mechanical-manufacturing-materials engineering perspective, fretting represents an important source of inefficiency for device design, particularly when we consider the issues of friction, inelasticity, wear, damage and fatigue cracking.

Fretting is a surface damage process generally associated with small amplitude relative motion between nominally static contact surfaces, e.g. clamped shaft-hub connections, splined shafts, dovetail connections, bolted joints, electrical connectors, interference fits. Fretting can be sub-divided into fretting wear and fretting fatigue. Fretting wear is associated with situations where there is no significant underlying fatigue stress, and the surface damage process is dominated by material removal or wear, leading to loss of function (e.g. looseness of fit) of the connection. Fretting fatigue is generally associated with an additional underlying fatigue (cyclic) stress or strain, leading to crack initiation from the surface, potentially with propagation and final fracture through a load-carrying member, e.g. shaft or housing, with more detrimental consequences. Fretting is commonly classified according to the sliding regime associated with the relative movement, which is a distinctive feature. Three sliding regimes are commonly identified in fretting, viz. gross slip, partial slip and mixed stick–slip. Gross slip occurs between two contacting bodies under normal (clamping) force when the tangential load exceeds limiting friction. Partial slip occurs when the tangential load is less than limiting friction and a stick region (typically in the centre of the contact) is maintained, with slip regions close to the

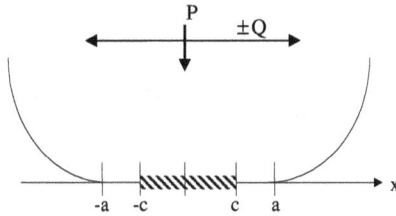

Fig. 1: Hertzian contact geometry (deformed) and definition of contact parameters, including contact semi-width, a, stick zone semi-width, c, normal load, P, and cyclic tangential load, Q. x is the independent horizontal coordinate parallel to the (flat) substrate contact surface.

edges of the contact, as shown in Fig. 1. The mixed slip regime occurs when a partial slip transforms into a gross slip contact, or vice-versa. This can occur for different reasons, including variation of friction conditions or breakdown or build-up of surface layers. Fretting wear and fretting fatigue are generally, but not exclusively, associated with the gross and partial slip regimes, respectively. Fretting fatigue cracks are generally found to initiate from within the partial slip zones.

Although it has been argued that no single fretting fatigue parameter exists, it is generally accepted that there is a set of primary variables which includes normal contact load (P), coefficient of friction (COF, μ) and slip amplitude (δ), e.g. see [1].

The challenge in fretting fatigue life prediction is to understand how the surface damaging mechanism affects initiation life. In conventional fatigue, crack initiation can account for up to 90% of the fatigue life, whereas in fretting fatigue this figure can reduce to as little as 5% [1]. The types of laboratory contact geometries employed for fretting fatigue testing, which generally consist of a pad or pair of pads clamped onto a plain fatigue specimen (Fig. 2), can be classified into complete and incomplete types, depending on whether the pad contact surface is nominally in complete contact with the fatigue specimen surface or not. For example, a flat-foot bridge type pad, similar to a punch-on-flat geometry, is classified as complete contact, whereas cylindrical and spherical pads (i.e. Hertzian) and flat pads with rounded edges are classified as incomplete (Fig. 1). Complete contact geometries have associated sharp contact edges.

Laboratory testing normally employs a fixed clamping load with a cyclic tangential load on the pads, along with a cyclic substrate stress on the plain fatigue specimen, which may either be in-phase or out-of-phase with the tangential loading. Generally speaking, the fretting fatigue behaviour of

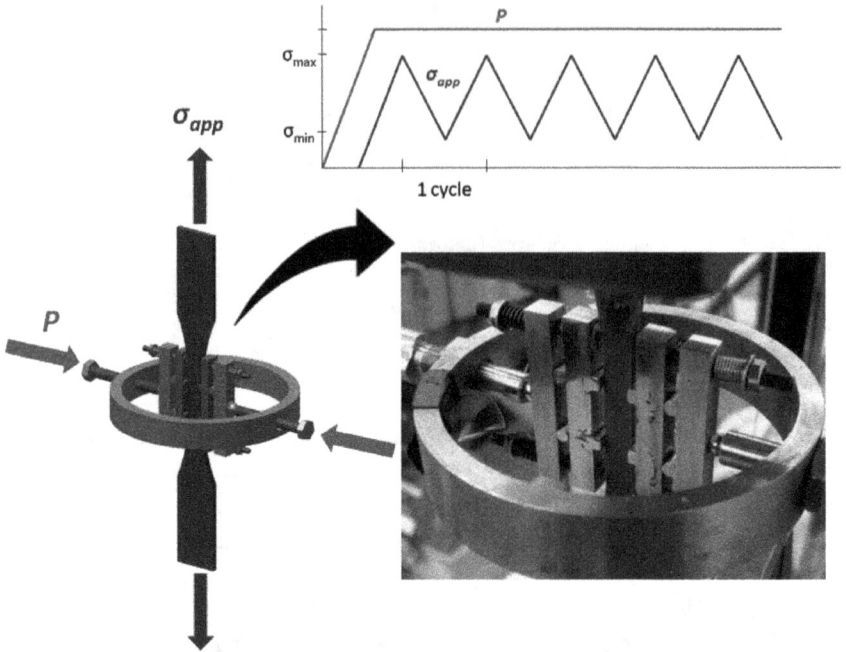

Fig. 2: Fretting fatigue bridge test arrangement for 316 stainless steel with bridge contact pair of pads clamped onto fatigue test specimen [2]. Also shown is a typical assumed loading history for normal load P and cyclic (fatigue) substrate load, σ_{app}.

each type of contact is different, due to the differences between the contact stress distributions and the slip distributions along the contact interface. For incomplete fretting geometries of the Hertzian type, the peak pressure under normal loading typically occurs at the centre of contact and reduces elliptically towards the edges of contact. For flat pad, rounded-edge cases, the peak pressure may not occur at the centre but will occur away from the edges. Figure 3 shows a typical relationship between fatigue life and wear rate as a function of slip amplitude for an incomplete contact. This relationship has been interpreted as the initiating fretting fatigue cracks being worn away with increasing slip amplitude [3]. There is generally a range of slip amplitude within which fretting cracks are initiated but do not get worn away so that fatigue life is significantly reduced in this range. Fretting cracks are generally found to initiate within the slip regions and the stick–slip interface is often considered a critical location. In contrast, for complete contacts, a steady-state partial slip regime is not theoretically possible, so that cyclic tangential loading leads to either complete stick or

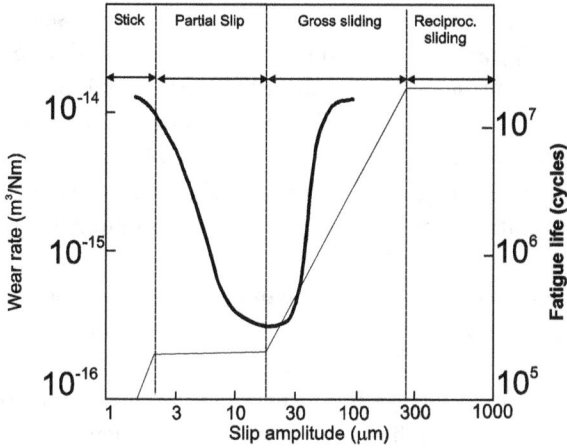

Fig. 3: Typical relationship between fatigue life and wear rate as a function of slip amplitude (after Vingsbo and Soderberg [3]), including the various slip regimes of stick, partial slip, gross slip (sliding) and reciprocating sliding.

complete sliding. In either case, complete contact fretting fatigue cracks generally occur at the sharp contact edges, where there is a theoretical stress singularity.

One of the major challenges for fretting fatigue design is the bridging of the gap between idealised (and simplified) laboratory tests, including associated theoretical and computer models, typically constrained to simplified loading conditions, and the need for real-life solutions, which relate more directly to the *in-situ* loading, environmental and other relevant conditions (e.g. geometrical constraints) of target machine or structural components. Clearly, the latter should be the application of all engineering research and yet often the laboratory context appears to become an end in itself. Similarly, there is a pressing need for computational modelling to make a greater impact in design, in this context, in design for fretting wear-fatigue. The difficulty with computational modelling for design is the significant times required for (i) development, calibration and validation, and (ii) parametric (sensitivity) studies for design variables. These problems are particularly exacerbated for nonlinear applications, e.g. frictional contact, plasticity, which in turn are compounded by uncertainties relating to nonlinear material properties (e.g. COF, wear coefficient) and model parameters.

The thesis of the present chapter is that the key step forward for fretting is (i) to recognise, on the one hand, the importance of spatial and temporal (field) distributions of key multiaxial surface and sub-surface parameters, rather than try to reduce down to a single parameter or set of simple variables such as (P, μ, δ), and, (ii) on the other hand, to develop experimental tests which represent these spatial-temporal distributions. This is, of course, entirely feasible, using modern nonlinear computational techniques, in combination with experimental and theoretical advances in fretting.

The chapter describes experiences and challenges relating to the application of computational modelling to the design of simple but representative tests for fretting fatigue and wear of aeroengine spline couplings.

The following sections of this chapter discuss (i) background on the industrial application, (ii) the general topic of design for fretting, (iii) fretting fatigue testing of a laboratory-scale spline coupling, (iv) application of computational methods (finite element) to fretting fatigue of spline couplings, (v) design of simplified representative specimen tests for fretting in splines, and (vi) simple parameters for prediction of fretting fatigue. Finally, there is a section on conclusions.

2. Industrial Application

The demonstrator application here is to spline couplings, especially the type of couplings employed in gas turbine aeroengines, to connect the low-pressure (LP) turbine and compressor shafts in triple-spool engines. These aeroengine spline couplings are designed for torque and axial load transmission, and experience wide variations in torque between take-off, cruise and landing conditions. However, the couplings also experience rotating bending moments due to gyroscopic effects during flight manoeuvres and engine carcass deformations. A small-scale spline testing facility was constructed at the University of Nottingham (UK), as part of the Rolls-Royce University Technology Centre (UTC) for Gas Turbine Transmissions, to improve the understanding of fatigue behaviour under such complex in-service loading conditions. The specimens were manufactured from a high strength CrMoV steel using the manufacturing route employed for service aeroengine shafts. They have 18 helical teeth on a reduced diameter but retain the representative tooth dimensions of real splines. Figure 4 shows the spline geometry, together with definition of the contact axial length coordinate z, the tooth flank contact width coordinate x and the maximum possible contact dimensions a_1 and a_2. More detailed information has been presented elsewhere [4].

An important feature of these spline couplings is the barrelled profile on the external spline. "Barrelling" refers to modification or optimisation of the thickness of the external spline teeth with respect to axial position along

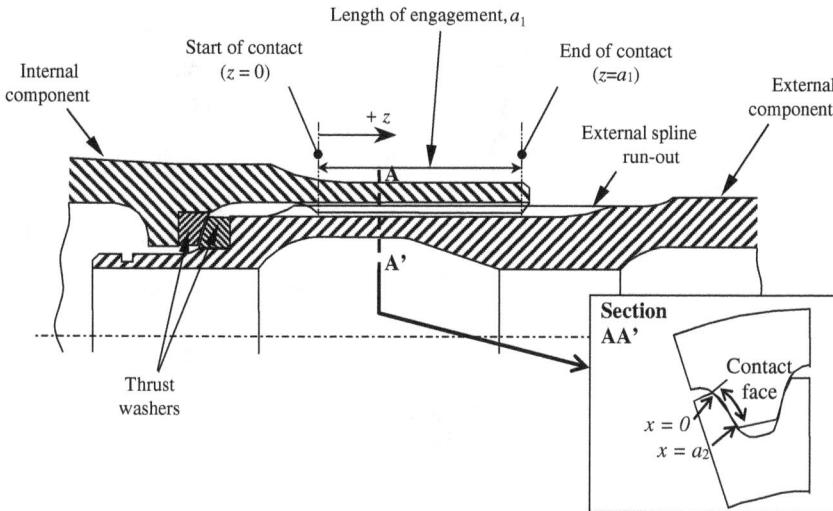

Fig. 4: Schematic half-section of the small-scale aeroengine splined coupling connecting turbine and compressor shafts showing important geometrical features, as well as tooth flank contact surface coordinate x and axial length of engagement coordinate z.

the shaft to create a more uniform axial contact pressure distribution (with reduced end-of-contact stress peaks). The profile is implemented by moving the spline cutter in towards the shaft axis (y_h in Fig. 5(a)), according to the required reduction in tooth thickness at each axial section. This, in effect, moves the external tooth involute profile inwards and at a certain angle. As a result, the new external tooth will no longer conform nominally to the profile of the internal spline tooth, as shown in Fig. 5(b). The schematic also shows that because the gap between the two mating teeth is larger at the external tooth tip, contact begins close to the root as the spline is loaded. The axial profile modification is such that the barrelling is zero near the axial centre of engagement and increases gradually (hence reducing tooth thickness) towards the $z = 0$, a ends of engagement (see [5] for a more detailed description).

The design of barrelled, helical spline couplings is intended to produce a near-uniform axial contact pressure distribution under the application of torque and to avoid the end-of-engagement peaks inherent in non-barrelled splines. The barrelled, helical spline coupling is thus an excellent example of a real-world, complex engineering contact geometry which does not readily lend itself to direct correspondence with the traditional, idealised fretting contact geometries, either incomplete, such as sphere- or cylinder-on-flat,

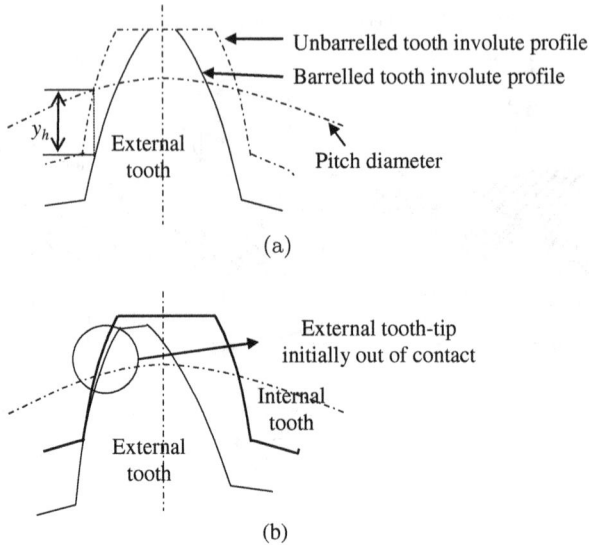

Fig. 5:   Schematic of (a) the result of cutter displacement, y_h in changing the external tooth profile; (b) the effects of barrelling on contact between the external and internal spline teeth. Drawing not to scale and barrelling exaggerated for clarity [6].

rounded punch-on-flat, or complete, such as punch-on-flat, typical of fretting laboratory research. While the transverse tooth flank (x) contact is somewhat close to a punch-on-flat type of contact, the axial (z) contact is incomplete and therefore more akin to a sphere on flat. The additional complexities of the effects of deformation, including bending stresses due to spline-tooth, cantilever-type geometry (similar to gear tooth bending) and the primary association of tooth-to-tooth contact (normal and tangential) tractions with shaft torque transfer, as well as the significant complexities of the inside diameter of the shafts (for more efficient and effective axial torque transfer [7]) lead to an onerous fretting fatigue problem, which is indeed extremely complicated to analyse.

The design of such a component against fretting fatigue is even more challenging due to the requirement for a test method to facilitate rapid evaluation of the many variables involved. For example, a key issue in fretting design generally is assessment of material and surface treatment requirements. The spline coupling is (obviously) manufactured from the main-shaft material, e.g. high strength CrMoV steel for aeroengine material. Non-matching (hard–soft) contact surface materials are commonly found to give superior performance. Nitriding has been adopted, for example, on one surface, with associated benefits in terms of surface hardness and beneficial (compressive) residual stresses. Another issue of key importance to fretting is environment, including surface lubrication. In an aeroengine main-shaft environment, for example, there is typically no active lubrication system for couplings, but there may be a mist of jet fuel, which can act as a lubricant. But clearly, this environment significantly affects evolution of COF with fretting cycles.

The spline contact is inherently three-dimensional (3D) with a complex 3D contact patch, e.g. see [8]. Consequently, this makes the wear-fatigue interaction and modelling of it significantly more complex than Hertzian or punch-on-flat fretting geometries, e.g. as studied by Madge *et al.* [9, 10].

3. Design for Fretting

3.1. *Fretting maps*

The concept of fretting maps was developed by Vingsbo and co-workers [3] to help characterise material behaviour and component running conditions with respect to fretting behaviour. Running condition fretting maps (RCFMs, Fig. 6) show plots of loads against applied stroke for graphical demarcation of the different sliding regimes. Material response fretting maps (MRFMs, Fig. 6(c)) show plots of stress against applied stroke for graphical demarcation of zones of no-damage, cracking and particle detachment (wear). Vincent *et al.* [11] have recommended the following steps to interpret fretting test information re-arranged into fretting maps and extrapolate to industrial problems or specific applications and devices:

1. Identify as closely as possible the contact operating conditions of the industrial component or application.
2. Situate them relative to the test results presented in the fretting maps.
3. Identify failure mechanisms, rather than try to obtain life reduction factors.
4. Attempt correlation between prototype failure information and laboratory tests.

The fretting maps approach (i) is based on experimental characterisation across the load space of fretting design variables(e.g. normal load, stroke), (ii) can be used for identification of optimal running conditions or material response assessment, and (iii) provides a powerful mechanism for systematic fretting wear-fatigue testing. However, it is not a design tool *per se*.

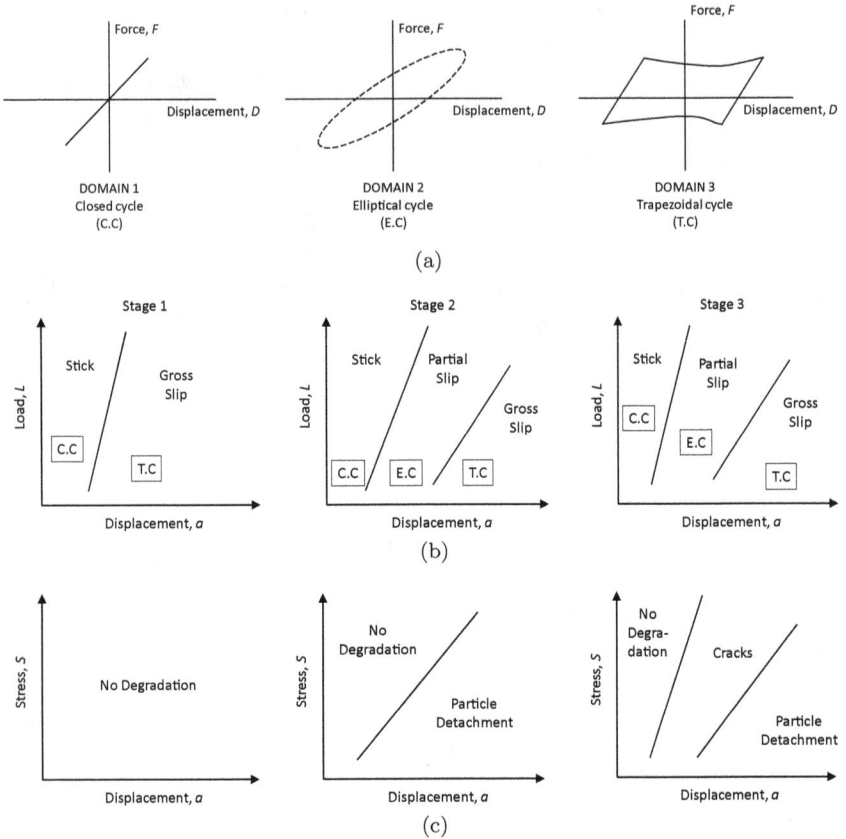

Fig. 6:   Fretting force–displacement loops and fretting maps [11]. (a) Characteristic fretting force–displacement loops; (b) running condition fretting maps (L = load, a = displacement); and (c) material response fretting maps (S = stress, a = displacement).

3.2. *Analytical methods*

Numerous authors have developed analytical solid mechanics solutions for contact problems using the theory of elasticity and mathematical solutions techniques, e.g. [12]. Hills and Nowell have presented an excellent text on the mechanics of fretting fatigue, with specific focus on such analytical techniques [13]. For complete fretting contacts, such as sharp-edged pads, Giannakopoulos *et al.* [14] have explored the analogy between the contact stress singularity at the contact edges and the stress singularity at a crack-tip, as shown in Fig. 7. A fracture mechanics based fretting fatigue life calculation methodology for simple contacts was presented, using an

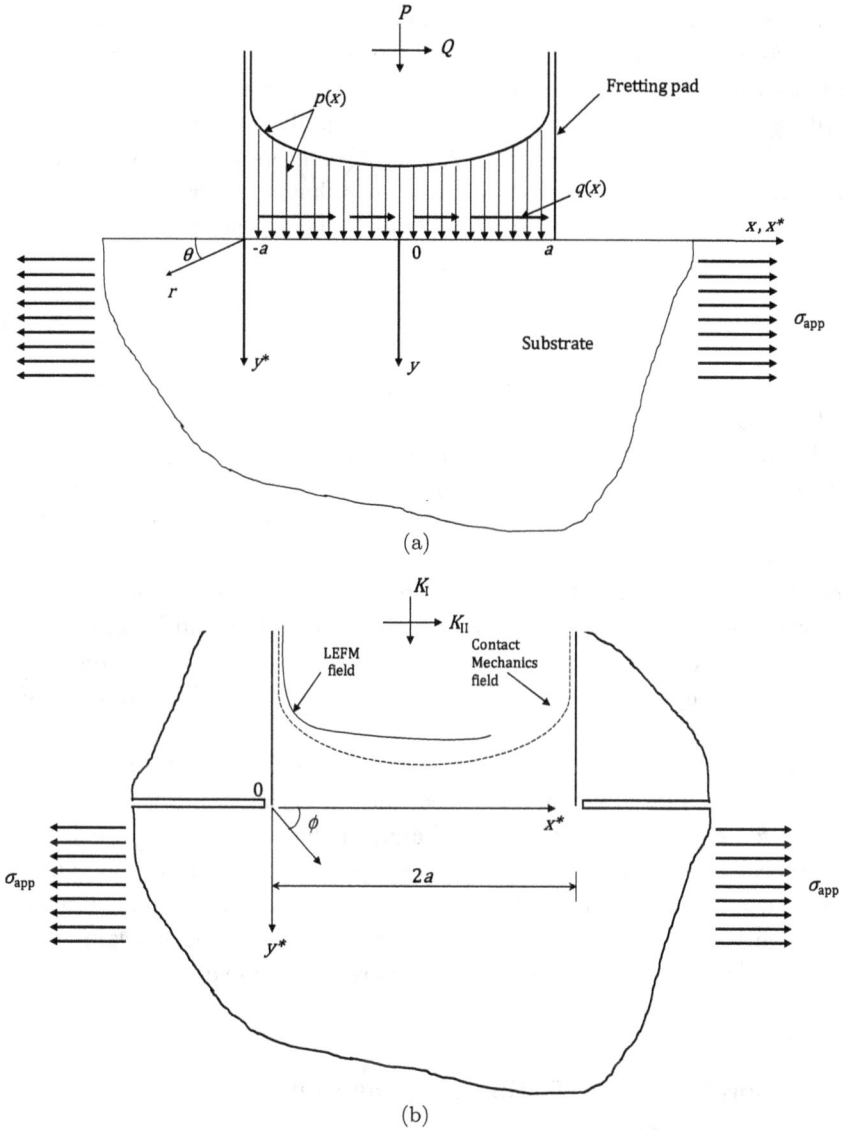

(a)

(b)

Fig. 7: Analogy between (a) punch-on-flat fretting contact with substrate stress, σ_{app}; (b) double-edge cracked plate specimen subjected to Mode I and Mode II stress intensity factors [14].

analogy between fracture mechanics and contact mechanics, to furnish the stresses for evaluation of fatigue crack growth, assuming an initial fretting-induced flaw size and direction from the "imaginary" (analogy) crack-tip. Hills and co-workers [5, 16] have explored a similar approach. The use of a generalised stress intensity factor, K^*, has been proposed, for characterising the stress distribution near the contact edge, having shown that the stress singularity associated with a sliding sharp-edged contact is given by the following expression:

$$\sigma_{ij} = K^* r^{\lambda-1} f_{ij}(\theta), \tag{1}$$

where λ is an eigenvalue ($-1 < \lambda - 1 < 0$) which is a function of contact edge angle (for cases where the contact edge is not perpendicular to the substrate) and coefficient of friction, r, θ are angular coordinates and the $f_{ij}(\theta)$ are defined in [16]. The concept of a process zone, where crack nucleation occurs, is introduced, with the objective of pursuing an analogy with fracture mechanics, whereby K^* can be used to map fretting fatigue crack nucleation predictions from simple test geometries to more complex geometries, provided the process zone is sufficiently small relative to the singularity-dominated region. The mapping of the simple test results to the more complex geometries will rely on matching of K^* values and slip displacements.

Although such methods require highly specialised expertise to develop, the key potential advantage is speed, once programmed into stand-alone code, for example. On the other hand, analytical methods are limited to specific, idealised contact geometries (punch-on-flat, Hertzian contact, etc.), which is a significant disadvantage for design, where contact geometry changes are of direct interest. Furthermore, these methods cannot normally include nonlinear effects, such as plasticity, large deformations, finite sliding.

4. Laboratory-Scale Testing of Spline Coupling

A programme of fatigue tests was conducted at the Rolls-Royce UTC for Gas Turbine Transmissions at the University of Nottingham [17] to simulate the key life-limiting loading experienced by splines during a typical civil flight envelope consisting of the combined major and minor cycle loading sequences as illustrated in Fig. 8. Each major cycle involves the application of a mean torque and axial load ramped up and held constant while 500 minor cycles are superimposed. Each minor cycle consists

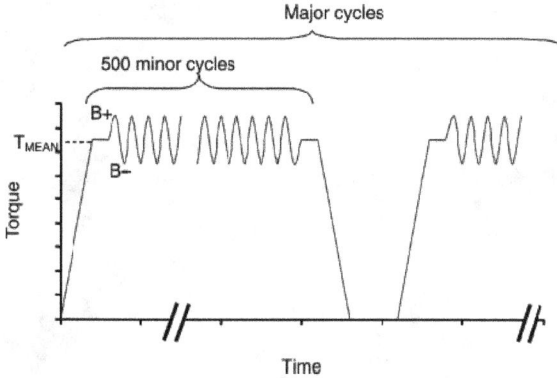

Fig. 8:   Schematic representation of major (torque and axial load) and minor (superimposed rotating bending moment and fluctuating torque) loading cycles, for spline fatigue testing.

Table 1:   Spline coupling test matrix to investigate effect of rotating bending moment (minor cycles).

Parameter	Test 1	Test 2	Test 3
Rotating bending moment	B_n	$2.22B_n$	$3B_n$
Test life to crack	$>1.7 \times 10^7$	6.01×10^6	7.45×10^5
Test crack location $(x/a_2, z/a_1)$	No crack	(0–1, 0.7–0.94)	(0–1, 0.7–0.87)
FE wear model life to crack	2.94×10^7	2.64×10^7	1.65×10^{6a}
FE wear model crack location $(x/a_2, z/a_1)$	(2.0,1.0)	(2.0, 1.0)	(1.0, 0.8)
FE D_{fret} life to crack	2.2×10^7	3.66×10^6	1.35×10^6
FE D_{fret} crack location $(x/a_2, z/a_1)$	(2.0,1.0)	(0.68, 0.97)	(0.68, 1.0)

Note: [a]For Test 3, a partial slip wear distribution is assumed, based on measured spline wear.

predominantly of a rotating bending moment of constant magnitude at a frequency of 5 Hz. Table 1 summarises the tests used to explore the role of rotating bending moment on fretting fatigue in the reduced-scale splines; the table also shows some FE-based predictions of fatigue and fretting fatigue locations and cycles, as described in more detail in what follows (see Section 5.2).

The details of spline test results have been reported in [17, 18]. It is found that in all cases cracking was confined to the externally splined shaft and the mode of loading affected the mode of failure. Minor cycle rotating

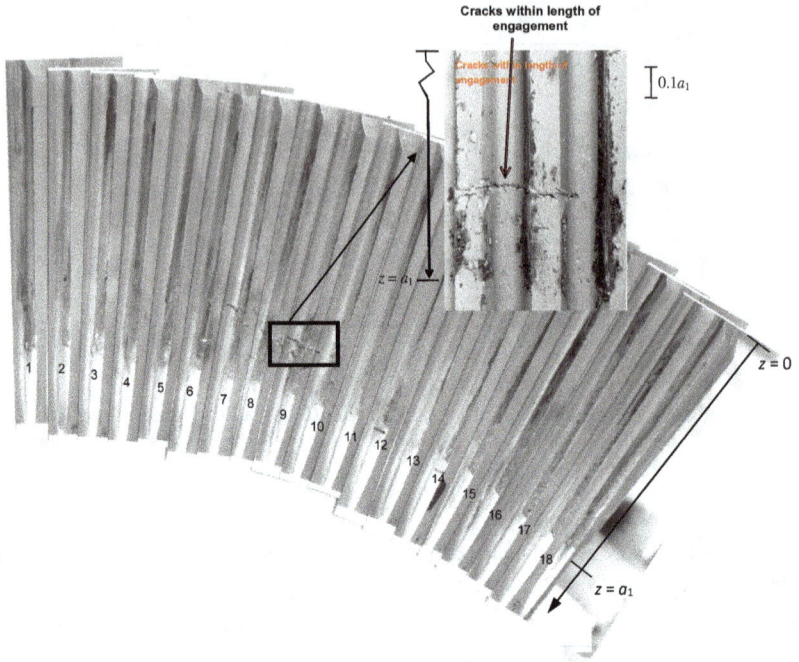

Fig. 9:　Photographic collage of spline external tooth flanks from Test 3 showing the crack path between teeth six and nine [18].

bending moments resulted in fretting fatigue cracks within the length of engagement and under the spline tooth contact regions, due predominantly to the increased relative slip associated with the bending moments.

Test 1, the design load case, showed no discernible cracks after 1.7×10^7 minor cycles. Tests 2 and 3 have bending loads of $2.22B_n$ and $3B_n$, respectively (B_n is design bending moment). These tests showed that life reduces significantly with increasing bending moment. Furthermore, the data shows that the location of cracking moves away from the $z = a_1$ (end of engagement) nominal contact edge with increasing moment to under the nominal contact length ($z < a_1$). It is interesting to point out that no cracking of teeth occurs at the $z = a_1$ end of engagement position, where theoretical stress peaks are expected, but wear occurs both in the radial direction and for a short distance along the axial direction (from $z = 0.9a_1$ to a_1), as shown in Fig. 9. The location of crack initiation transversely along the tooth involute, x/a_2, is between 0 and 1.0, i.e. within the nominal contact area.

5. Computational Modelling of Fretting in Splines

5.1. *3D FE model of spline coupling frictional contact*

Previous studies, e.g. [17], have shown that a one-tooth, cyclic symmetry FE model simulating one 20° segment of the coupling, as shown in Fig. 10, can be used for cyclically symmetric loading conditions, such as torque and axial load. This simplification has previously facilitated a high level of mesh refinement and thus detailed stress, strain and contact variable results, so that the fatigue behaviour of the coupling under the major cycle (torque and axial) loading can be accurately captured. However, for non-symmetric loading, i.e. rotating bending moment with in-phase fluctuating torque, which is clearly the case for fretting fatigue in splines, it is necessary to model the full 18-tooth 360° geometry. The need for accurate stress and strain predictions would significantly increase the demand on mesh refinement at the cost of computational resources. A localised mesh refinement technique using multi-point constraints (MPCs) was therefore adopted [19]. The full FE model is shown in Fig. 11; eight-node linear brick elements are used throughout. Figure 12 shows a cross-section of the external spline model showing the tooth numbering convention used and the level of mesh refinement, especially on the top-dead-centre tooth (i.e. Tooth 18). It can be seen that this approach has permitted the use of a fine FE mesh on Tooth 18 with a very much coarser mesh on all other teeth.

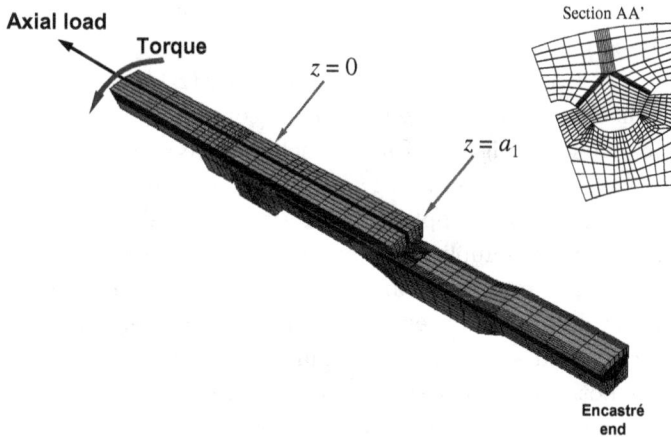

Fig. 10: One-tooth cyclic symmetry FE model of the spline coupling contact including turbine and compressor shaft splines, using coarse mesh for efficient wear simulation predictions.

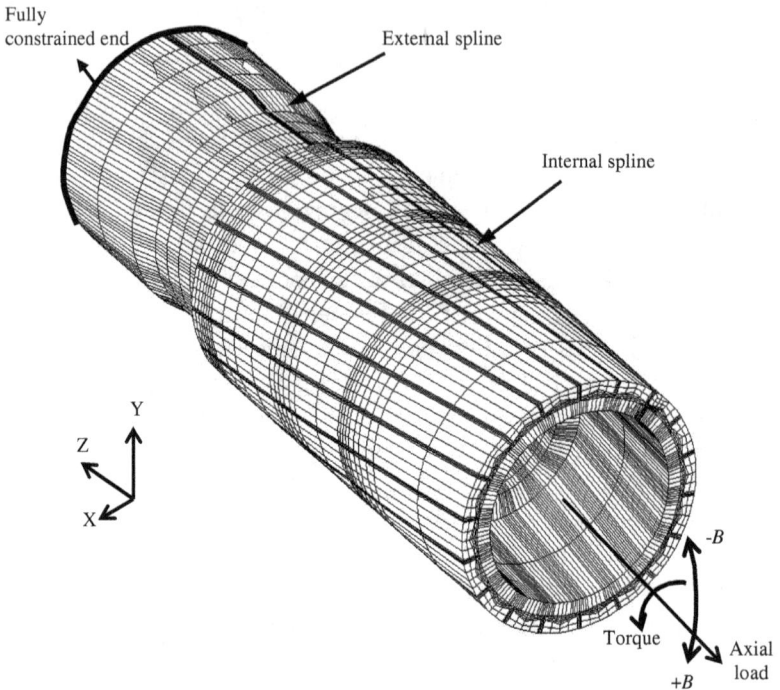

Fig. 11:   Full 360° (18-tooth) FE model of the spline coupling showing constraints on externally splined (turbine) shaft and loads on internally splined (compressor) shaft [22].

Negligible plastic deformation is predicted to occur for the range of load cases analysed here.

In the FE model, torque, bending moment (B) and axial loads are applied at the rear end of the internal spline (compressor shaft) in the direction of the arrows of Fig. 11. Torque fluctuates between 105% T_{mean} and 95% T_{mean} during minor cycles (where T_{mean} is the mean torque). A rotating bending moment has been applied, as described in detail by Ding *et al.* [20]; the bending moment is applied at a constant magnitude about a transverse axis, which rotates around the axis of the shaft. Six load steps (two complete minor cycles) are sufficient to capture the steady-state behaviour of stresses, strains and slips during the minor-cycle loading. Tooth 18 (top-dead-centre tooth), experiences maximum tensile stresses from the rotating bending moment $(+B)$, at the same time that it experiences the maximum total torque, T_{max}.

In this study, the basic Coulomb friction model with isotropic friction is employed within the contact regions. Frictional contact conditions are determined via the penalty method with an allowable elastic slip of

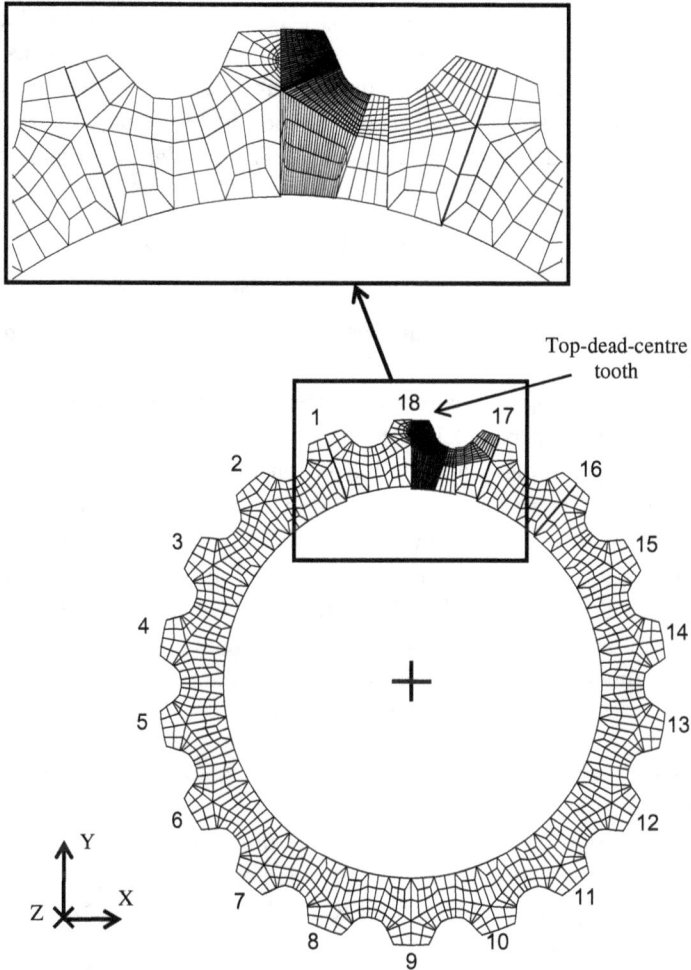

Fig. 12: Cross-section through the 360° (18-tooth) FE model of the externally splined shaft showing the tooth numbering scheme and the localised mesh refinement on Tooth 18 (z-axis directed out of page) [19].

$\delta/a_2 = 2.8 \times 10^{-5}$. Based on cylinder-on-flat tests for a representative range of loads and strokes [21], a coefficient of friction (COF) of 0.3 is assumed as a representative average value over 100,000 cycles.

5.2. *Modelling wear in spline coupling*

Accurate representation of the contact stress and strain fields in fretting requires accurate representation of the contact geometry, as well as the contact slip regime, including especially stick–slip interfaces. Modelling of material removal due to wear is thus essential for fretting fatigue, as it predicts not only the extent of wear damage, but also the concomitant change in fatigue-pertinent stresses and strains. A modified form of Archard's equation [23, 24] is used here to predict the local wear depth increment for a given point within the contact:

$$\mathrm{d}h = kp\mathrm{d}s, \tag{2}$$

where h is the wear depth (mm), k is the dimensional wear coefficient defined as the wear per unit slip per unit contact pressure, p is the local contact pressure and s is contact slip.

Ding [21] developed a computer programme to implement this equation incrementally in parallel with the ABAQUS finite element code and the results of this implementation have been validated against cylinder-on-flat fretting wear tests for the spline material combination employed here, over a range of stroke and load combinations [23]. More recent work by Madge *et al.* [9] has implemented Eq. (2) via an adaptive meshing user subroutine called UMESHMOTION, within ABAQUS; in this case, the application was to the prediction of fretting fatigue life of titanium fretting fatigue specimens. The user subroutine implementation is more accurate and efficient than that of [23] in the following aspects:

- The spatial adjustment of the contact nodes is achieved within an adaptive meshing framework.
- One implication of this is that, whereas previously the wear depth was calculated using the contact pressure for normal loading only and with an average relative slip over the tangential load cycle [23], now the incremental wear depth is calculated for each increment of the tangential load cycle (fretting cycle) using the associated instantaneous values of contact pressure and relative slip.
- Whereas previously a separate analysis was needed for each cycle-jump, with an attendant unloading and re-loading of the normal load, the new implementation circumvents this, since the geometrical modification is effected within the FE code itself. This is more consistent with the actual experimental situation.
- Small cycle jumps give finer discretisation of the time domain helping to maintain stability and accuracy; larger jumps decrease the computational expense. Since the present implementation modifies the geometry incrementally throughout the tangential cycle, it is more stable and hence much larger cycle jumps can be made.

(a)

(b)

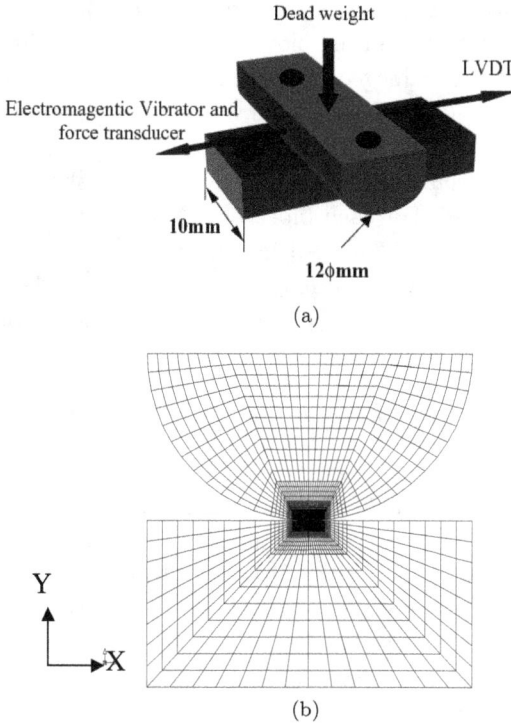

Fig. 13: (a) Schematic representation of cylinder-on-flat fretting wear test geometry; (b) 2D finite element contact model for wear simulation.

However, one disadvantage of the user subroutine approach is that all of the fretting cycle wear incrementation data needs to be stored in memory while the single analysis is running. This leads to limitations to the model size and incrementation discretisation, which in some cases, can become prohibitive.

The results of the two approaches have been shown to be indistinguishable for cylinder-on-flat (Fig. 13(a)) fretting wear of the spline high-strength steel (under partial and gross slip conditions), using the 2D FE model of Fig. 13(b) [20]; other details of the FE model and fretting tests are available in [23]. A wear coefficient of $5.0 \times 10^{-8}\,\mathrm{MPa}^{-1}$, which is representative of the high-strength steel up to 18,000 fretting wear cycles [23], was employed. The results have been previously validated against the fretting scar test data (gross slip) for a range of contact loads. From the point of view of application to the complex spline coupling wear simulation, the user subroutine is preferable.

The spline wear modelling implementation is within UMESHMOTION. A cycle-jumping technique is employed whereby the required total number of wear cycles N_T is achieved by n wear blocks, each corresponding to a block of ΔN wear cycles. The feasible values of ΔN are determined by the stability of the subsequent contact pressure and slip calculations. Generally, as the pressure variation across the contact diminishes, higher values of ΔN can be tolerated. For each block of ΔN wear cycles, the normal and tangential loading cycle is applied incrementally for $m = 1$ to m_{inc} (m is increment number). A value of $m_{\text{inc}} = 20$ is used here. Following Eq. (2), the wear depth increment $\Delta h_{i,m}$ at each contact node i for the mth load increment is calculated by:

$$\Delta h_{i,m} = k\,\Delta N\, p_i\, \Delta s_i = k\,\Delta N\, p_{i,m}\, \Delta s_{i,m}, \tag{3}$$

where $p_{i,m}$ and $\Delta s_{i,m}$ are contact pressure and (resultant) slip increment, respectively, at contact node i for increment m.

Equation (3) provides the magnitude of the wear increment $\Delta h_{i,m}$ for each nodal position, but not the direction. To obtain this, the directions of the local slip and contact pressure are required. ABAQUS computes the local slip components in the local slip plane, which is transverse to the local outwards surface normal. The local contact pressure is also in the direction of the local outwards surface normal. Thus, at each "worn node", the wear depth should be proportional to the product of the local resultant in-plane slip and the component of the contact pressure normal to the slip plane. As a result, "worn nodes" will be moved off their original circumferential orbits when the wear depth increments are applied (Fig. 14). To avoid a loss of cyclic symmetry, the wear direction in this study is assumed to be tangential to the original "orbits". The geometrical implementation is illustrated in Fig. 14, which shows two contact nodes (i and $i+1$) on a transverse section through an external tooth segment after increments $m-1$ and m. From Fig. 14 it can be readily shown that:

$$X_{i,m} = R_i \cos(\theta_{i,m-1} - \Delta\theta_{i,m}), \tag{4}$$

$$Y_{i,m} = R_i \sin(\theta_{i,m-1} - \Delta\theta_{i,m}), \tag{5}$$

where $\theta_{i,m-1}$ is equal to $\tan^{-1}(X_{i,m-1}, Y_{i,m-1})$ and $\Delta\theta_{i,m}$ is equal to $\Delta h_{i,m}/R_i$. The above sequence is repeated until the required total number of wear cycles N_T is reached. The simulation of incremental wear in a finite element model of a complex geometry like that of the spline is however a computationally intensive process, with typical simulations taking one and

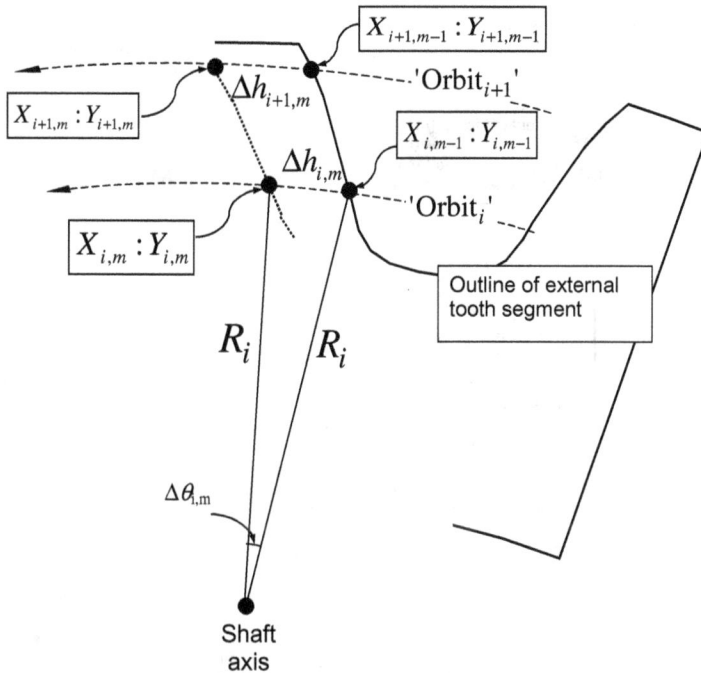

Fig. 14: Schematic transverse section through an external tooth segment showing the movement of contact nodes in response to step n and the displacement required to return node i to its "orbit" after this step. X and Y are global coordinates (not to scale).

half days for about 12,000 wear cycles on a typical personal computer. The procedure employed here is presented in Fig. 15(a).

Figure 16 shows predicted distributions and cyclic evolutions of wear depth over a range of axial and tooth flank positions after 12,000 major (mean torque and axial) cycles of spline loading. Table 2 shows a comparison of predicted and measured mean wear depths from two contrasting spline fatigue tests. Reasonable agreement is achieved considering the significant number of simplifications and assumptions required to arrive at a solution.

5.3. *Prediction of fretting fatigue in spline coupling with wear*

Madge and co-workers were the first to present a multiaxial fretting fatigue damage methodology incorporating the effects of material removal due to wear on evolution of contact stress–strain distributions, to successfully predict the experimentally measured effect of slip amplitude on fretting

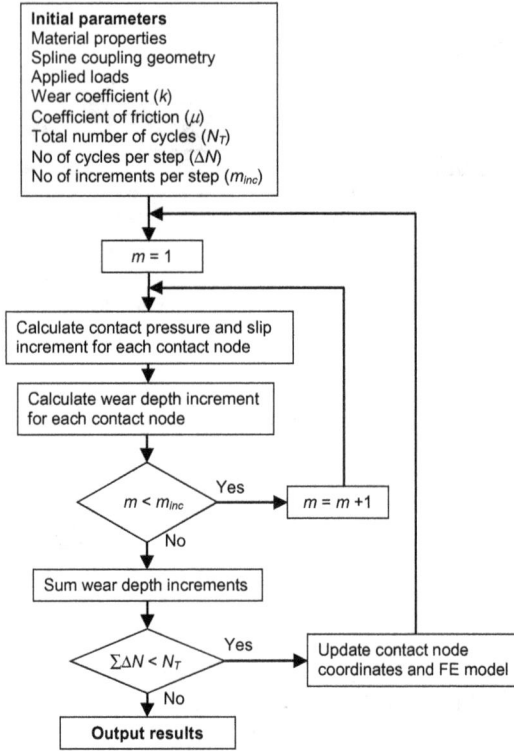

(a)

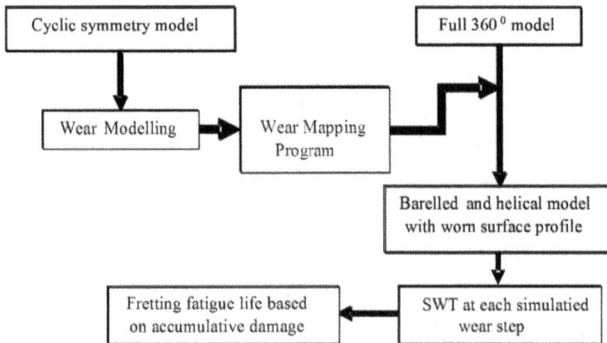

(b)

Fig. 15:   (a) Flowchart for wear simulation of spline coupling; (b) analysis procedure for critical-plane fatigue damage parameter SWT with evolving contact geometry.

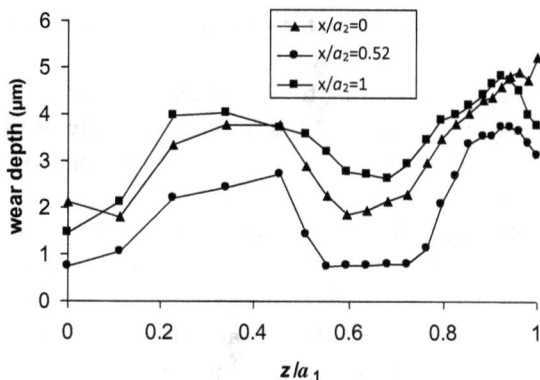

(a)

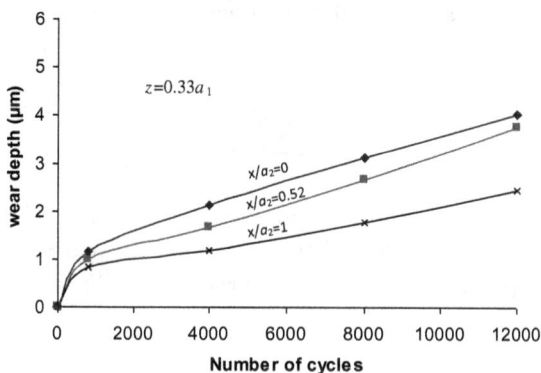

(b)

Fig. 16: (a) Predicted wear depth versus axial position; (b) predicted wear depth versus number of cycles at $z/a_1 = 0.33$ axial position, for different sample tooth flank positions after 12,000 spline major (torque and axial) load cycles.

Table 2: Comparison of predicted and measured mean wear depths for spline coupling under major (torque and axial) cyclic loading.

Test no.	No. of major cycles	Measured and predicted mean wear depths $(z/a_1 = 1)$ μm	
		Measured	Prediction
1	35,000	8	12[a]
2	12,000	5	4

Note: [a]Extrapolated value, based on 12,000 simulated cycles.

fatigue life [9, 25]. Specifically, this captured the "bath-tub" fatigue life trend of Fig. 3, whereby the peak stick–slip interface stresses are accentuated by partial-slip wear, leading to significantly increased fatigue damage, whereas the peak trailing edge stresses are beneficially redistributed over an ever-widening contact width and attenuated for gross-slip wear, leading to reduced fatigue damage and hence longer life. This work was based on 2D models of fretting fatigue test geometries, where the evolution of the stress–strain histories was simulated for each wear step and a linear damage accumulation law (i.e. Miner's rule [26]) was employed to accumulate incremental fatigue damage (due to wear-induced variation of stress and strain from one fretting cycle to the next) at each material point, to predict total life. Fretting fatigue failure at a material point is therefore defined to correspond to the total accumulated damage w reaching a value of 1, where w was defined as

$$\omega = \sum_{n=1}^{n=\frac{N_T}{\Delta N}} \frac{1}{N_{f,n}}, \tag{6}$$

where $N_{f,n}$ is the number of combined (major–minor) cycles to failure for the stress–strain state predicted to correspond to wear step n, calculated using a critical-plane implementation of the Smith–Watson–Topper (SWT) equation as follows:

$$\sigma_{\max}\Delta\varepsilon_a = \frac{(\sigma_f')^2}{E}(2N_f)^{2b} + \sigma_f'\,\varepsilon_f'(2N_f)^{b+c}, \tag{7}$$

where $\sigma_{\max}$ is the maximum normal stress on the critical plane, $\Delta\varepsilon_a$ is the maximum normal strain amplitude on the same plane, σ_f' and b are the fatigue strength coefficient and exponent, ε_f' and c are the fatigue ductility coefficient and exponent, E is Young's modulus and N_f is the number of cycles to failure, respectively. The SWT approach is based on physical observations that fatigue cracks initiate and grow within a material on certain planes, where the growth and orientation depends on the normal stresses and strains on these planes. A detailed procedure to determine the critical plane and the associated value of SWT can be found in Ref. [27], wherein it was shown that the critical-plane SWT approach predictions gave good correlation with measured plain fatigue lives for the present spline coupling.

Corresponding to a given wear step n, the number of combined cycles $N_{f,n}$ to failure is also predicted using Miner's rule. With 500 minor cycles per major cycle, it can be expressed as:

$$\frac{1}{N_{f,n}} = \frac{1}{N_{f,n}^{\text{maj}}} + \frac{500}{N_{f,n}^{\text{min}}}. \tag{8}$$

The major and minor cycle lives, $N_{f,n}^{\text{maj}}$ and $N_{f,n}^{\text{min}}$, are determined using the critical-plane SWT values for the major and minor cycle load cycles, respectively. Equations (6)–(8) can thus be employed to predict the number of cycles to failure ($\omega = 1$).

Accurate wear and fatigue prediction for any geometry using an FE approach requires as fine a mesh as possible, although compromise is always required between resolution/accuracy of predicted variables and run-time. Most analysis of fretting employs comparatively simple geometries and, hence, models. However, the requirement of simulating wear-fatigue in the spline contact introduces a number of complicating and competing factors as follows:

1. Given the use of Eq. (3) for geometry modification, relatively smooth predicted distributions of contact pressure and slip are desirable for accuracy and resolution, particularly in partial slip where mixed stick–slip occurs and for cases where contact geometry changes become discontinuous.
2. Incremental simulation of material removal requires a large number of nonlinear (frictional contact) increments, particularly for accurate resolution of tangential loading effects where transitions from stick to slip and vice-versa occur.
3. For accurate resolution of stresses, strains and fatigue parameters, within contact problems, with discontinuities in contact geometry, e.g. end of engagement, stick–slip interfaces, barrelling, a fine mesh is required, both at and below the surface.
4. For cyclically symmetric loading, a one-tooth model is sufficient, whereas for non-symmetric loading, a 360° model is required, with a fine mesh in the regions of interest, e.g. where significant changes in distributions of key fretting variables occur.
5. The geometrical complexities of the spline coupling itself, by definition, result in a large number of elements and nodes.
6. The complexities of the spline loading cycle, including frictional contact and plasticity effects, as well as the combined (major and minor) loading cycles, require a number of cycles, each with multiple increments, to ensure convergence.

The 18-tooth 360° model consists of about 150,000 elements, where a typical element dimension at the local mesh refinement region is about $0.028a_2$ in the x direction. It is not feasible from a run-time perspective (not to mention disk space) to use this model to simulate the wear resulting from, say, 12,000 cycles, with $\Delta N = 500$ cycles, at most. Hence, to achieve an acceptable computational overhead for wear simulation, a significantly coarser mesh is required. In the one-tooth coarse mesh spline model, as illustrated in Fig. 10, the average element dimension immediately below the contact region is $\sim 0.22a_2$ in the x direction and $0.55a_2$–$1.11a_2$ in the z direction. This is satisfactory from a wear simulation perspective but not from a fatigue life prediction perspective. Hence, a combined use of the cyclic symmetry model and the 360° model for the wear-fatigue predictions was implemented as depicted in the flowchart of Fig. 15(b). A number of implications follow from this:

1. The incremental wear simulation is limited to the one-tooth model. This is predicated on cyclic symmetry and therefore equal wear is assumed on all teeth, so that only major cycle wear, i.e. due to torque and axial load, is modelled incrementally.
2. The difference in mesh requirements for fatigue and wear analysis requires a technique to transfer the wear patterns obtained from the coarse mesh to the fine mesh model, to allow higher resolution damage accumulation with evolving contact geometry to be predicted.

In lieu of an incremental wear simulation for the 360° model, a "Kriging" surface interpolation approach [28] is used to investigate wear effects. The "Kriging" method belongs to the family of linear least squares estimation algorithms, which aims to estimate the value of an unknown real function f at a point A^*, given the values of the function at some other points $A_1, \ldots, A_n$. A Kriging estimator value $f(A^*)$ is a linear combination that may be written as

$$f(A^*) = \sum_{i=1}^{n} \lambda_i \, f(A_i). \tag{9}$$

The weights λ_i are solutions of a system of linear equations which are obtained by assuming that f is a sample-path of a random process $f(A^*)$, and that the prediction error

$$error\,(A^*) = F(A^*) - \sum_{i=1}^{n} \lambda_i \, f(A_i), \tag{10}$$

is to be minimised. A key benefit of this approach for this application is that data can be interpolated from a non-regularly spaced grid distribution of nodes. Hence, applying the Kriging interpolation approach to the predicted wear depth distribution from the cyclic symmetry model, the surface distribution of wear depth, for each wear step $n(n = 1, \frac{N_T}{\Delta N})$, can be obtained for the 18 teeth on the 360° model, with their various degrees of mesh refinement. The overall wear–life interaction procedure is depicted in the flowchart of Fig. 15(b). Key results from the FE wear-fatigue simulations are given in Figs. 17–21 and also in Table 1.

Some key points from the results of the FE wear-fatigue simulations of the spline coupling under combined major (torque and axial) and minor cycle (rotating bending and fluctuating torque) loading are as follows:

1. As shown in Figs. 17 and 18, the model predicts (i) gross slip, mainly in the tangential (transverse) x-direction, for major cycles and (ii) partial slip, mainly in the axial z-direction, for minor cycles. For the partial-slip case, slip increases with (i) axial distance, z, along the spline, towards the $z = a_1$ end, and (ii) increasing rotating moment.
2. The FE-predicted slip regions are broadly consistent with the locations of fretting fatigue cracks.
3. The tests show that fretting fatigue cracks (in the contact regions) are caused by bending moment overload, whereas major cycle (torque–axial) over-load causes spline root plain fatigue cracks.
4. The FE model critical-plane SWT distributions (Fig. 19), show the same effect of increasing moment as in the test, viz. increasing damage (reducing life) with increasing moment.
5. The no-wear (Fig. 19) and gross-slip wear FE models tend to over-predict life somewhat. The no-wear model predicts cracking at the $(x/a_2, z/a_2) = (1, 1)$ edge of contact, instead of the slip region, under the contact. The gross-slip wear model predicts that material removal causes plain fatigue cracking in the spline root for Test 2.
6. The partial-slip wear model (Figs. 20 and 21), only applied for Test 3 case, predicts cracking at $(x/a_2, z/a_2) = (1, 0.8)$ which is in agreement with the test and furthermore reduces the over-prediction of fretting fatigue life, to give much closer correlation with the test result.

In summary, the full 3D (multiaxial) distributions of cyclic stick–slip, stresses, strains, fatigue indicator parameter (in this case, critical-plane SWT), combined with the effects of both gross and partial-slip wear are

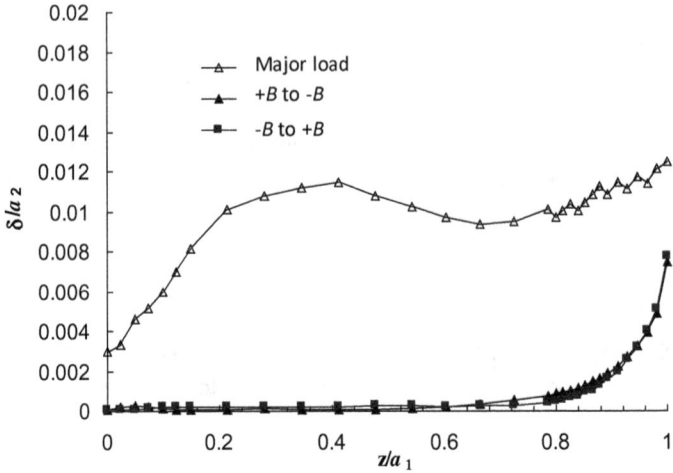

(a)

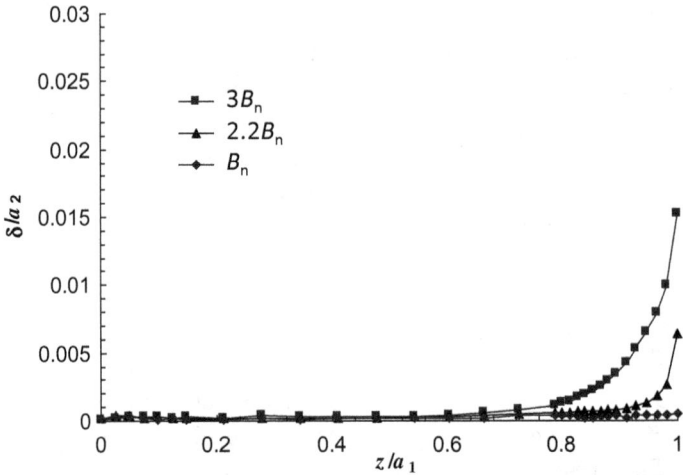

(b)

Fig. 17: (a) Stabilised resultant slip distributions versus axial position for major and minor cycle load of Test 3; (b) effect of rotating bending moment magnitude on axial distributions of resultant stabilised slip, both at $x/a_2 = 0.96$ tooth flank position.

critical to understanding the complexities of fretting fatigue in this helical, barrelled spline coupling under the combined effects of mean and fluctuating torque, axial loading and rotating bending moment. In particular, there are complex multiaxial, interactive effects between (i) gross slip due to major

Axial distance, z/a_1

| 1 | 0.9 | 0.8 | 0.7 | 0.6 | 0.5 | 0.4 | 0.3 | 0.2 | 0.1 | 0 |

Open

Slip

Distance from
tooth tip, x/a_2

(a)

Axial distance, z/a_1

| 1 | 0.9 | 0.8 | 0.7 | 0.6 | 0.5 | 0.4 | 0.3 | 0.2 | 0.1 | 0 |

Open

Slip

Slip

Stick zone

Distance from
tooth tip, x/a_2

(b)

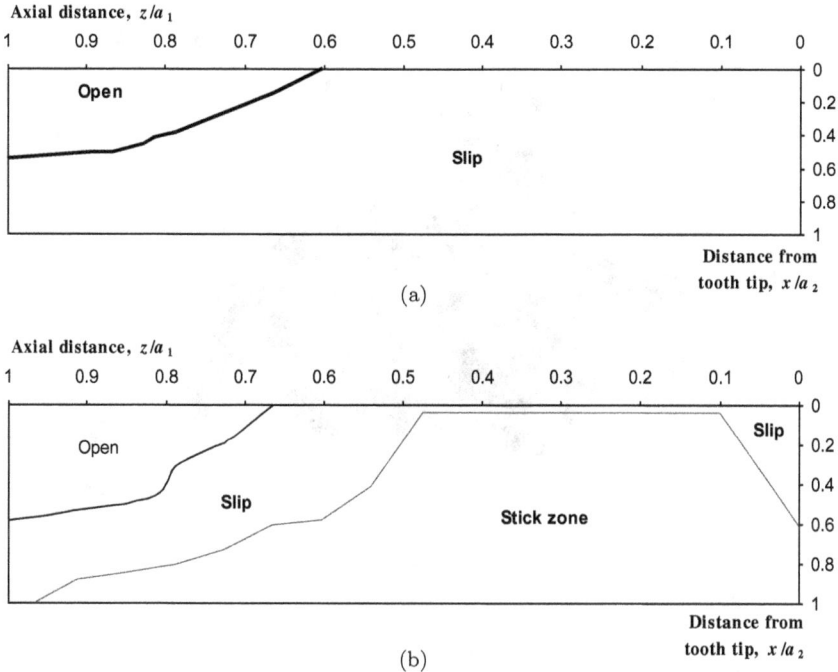

Fig. 18: Variation of Test 3 instantaneous contact status at: (a) major; (b) minor cycle.

(torque–axial) cycles, (ii) partial slip due to fluctuating torque with rotating bending moment, and (iii) associated multiaxial stress–strain distributions.

6. Representative Test Specimen Concept for Fretting in Splines

Figure 22 shows a representation of various levels of testing relevant to design for fretting of spline couplings. The present work is concerned with the development of simplified representative test specimens for fretting in splines. The most fundamental and common testing method for fretting is a simple cylinder-on-flat arrangement. This "cheap and cheerful" approach can be easily used for many tests and can provide important information about the representative tribological response of the relevant materials, such as evolution of friction and wear, as a function of normal load and applied stroke, and can mimic gross and partial slip to provide information on crack nucleation.

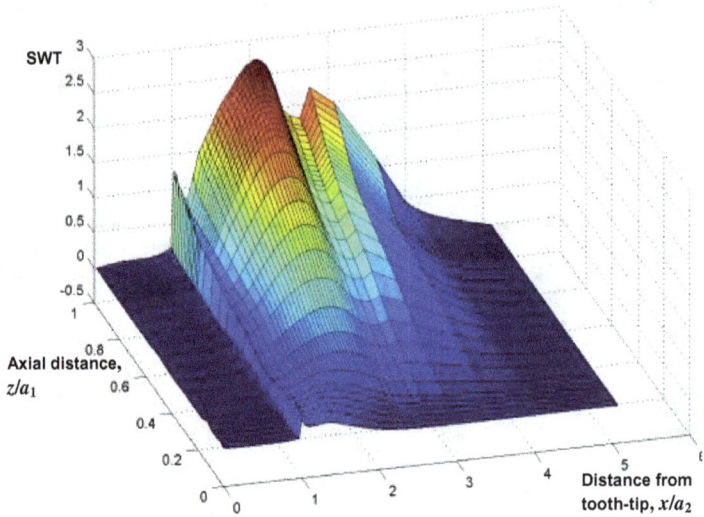

(a)

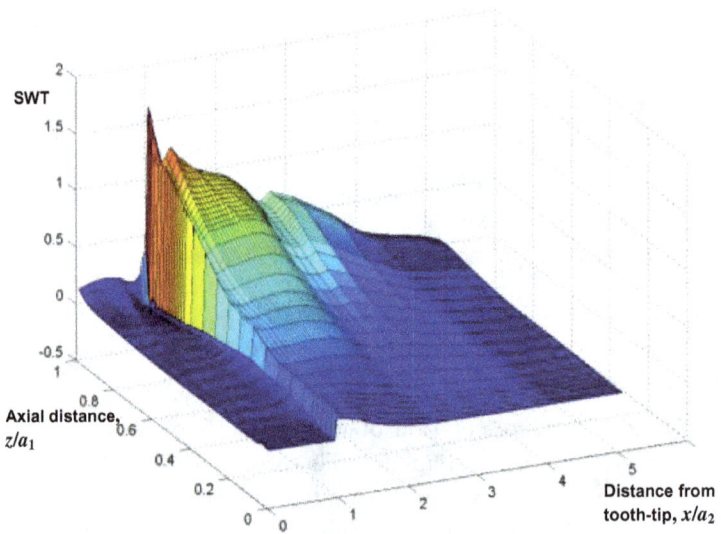

(b)

Fig. 19:   Predicted no-wear critical-plane SWT distributions over spline tooth (from tooth tip on Tooth 18 to adjacent tooth tip on Tooth 17, including spline fillets and root) for Test 3 loading conditions, for (a) major cycle loading; (b) minor cycle loading.

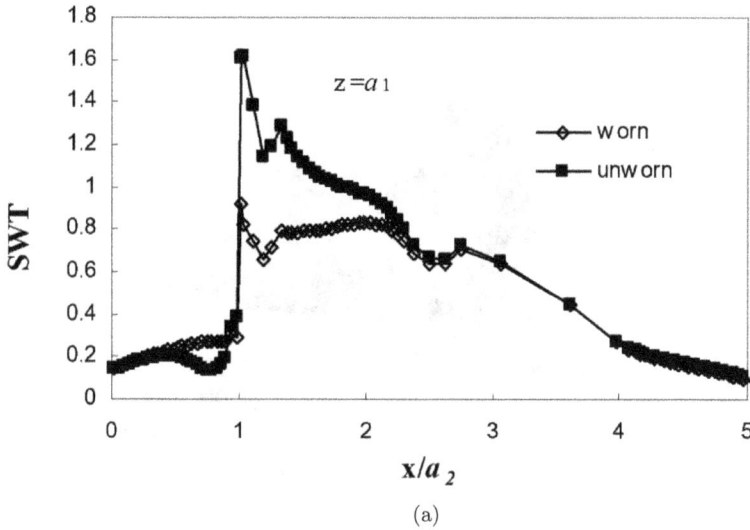

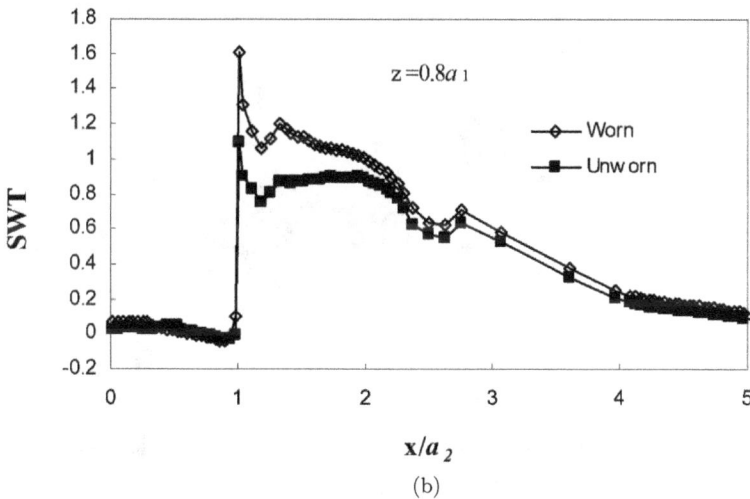

Fig. 20: Predicted effect of partial slip wear on minor-cycle critical-plane SWT distributions for Test 3 load case at: (a) $z/a_1 = 1$; (b) $z/a_1 = 0.8$.

However, this approach uses simplifying assumptions, is therefore somewhat limited and cannot, for example, represent: (i) the local complex contact geometry of a helical, barrelled spline coupling, (ii) the associated local and substrate distributions of substrate stresses and strains, (iii) the local complex multiaxial tribological behaviour, viz. gross-partial stick–slip regimes under combined torque, bending and axial loading. At the other

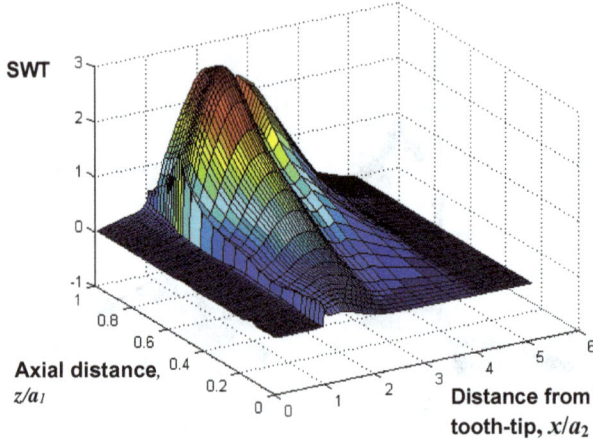

(a)

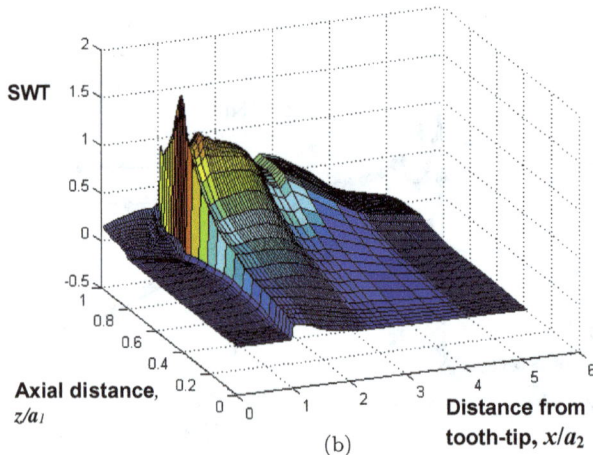

(b)

Fig. 21: Predicted (a) major; (b) minor cycle critical-plane SWT distributions for partial slip wear for Test 3 load conditions.

extreme, the full-scale, engine test is a much more expensive option and although it (arguably) contains the most complete representation of the in-service conditions for design, only a limited number of tests (perhaps even only one) per engine development cycle can be undertaken, so that it can really only be used as a check. Furthermore, it is not generally possible to derive component-specific design information from such tests, as attention will not typically be focused on a single component, like a spline coupling. The alternative of laboratory-scale spline tests, as discussed in Section 3,

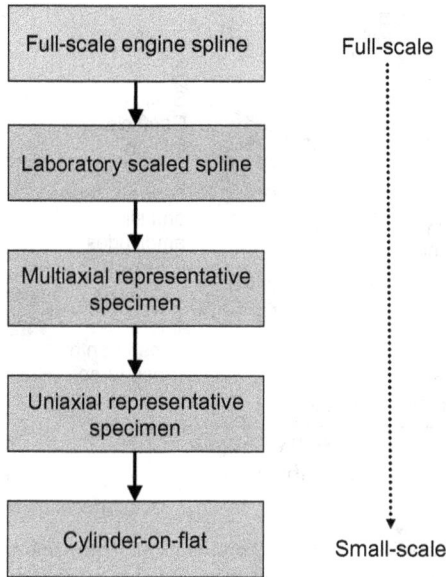

Fig. 22: Testing hierarchy relevant to design for fretting and fatigue of spline couplings [29].

is obviously extremely useful for validation and calibration of models, but is also overly expensive for design studies, e.g. evaluation of a new surface treatment option. Hence, there is a need for simplified representative test methods for fretting fatigue of complex, expensive couplings.

Figure 23 shows a combined experimental-computational methodology proposed to address the development of a representative test specimen for fretting in splines [30], shown in Fig. 24. This approach is referred to here as a uniaxial representative specimen (URS). It aimed at simulating the effects of major cycle (torque–axial) loading of the laboratory spline coupling [31]. The simplified URS approach can be used to successfully mimic the spline tooth interactions in a spline tooth coupling if finite element analysis is used to match the URS contact tractions and relative slip, during the design stage, to those of arbitrary positions along the spline coupling. It is not necessary to generate a detailed finite element model of all aspects of the URS, particularly the interfaces and associated fasteners, if empirical calibration of the URS compliance is employed. The observed trends in friction loop development for the URS tests are similar to those for cylinder-on-flat tests; however, by virtue of closely simulating the spline contact geometry and variables, the URS provides superior information on evolution of COF for

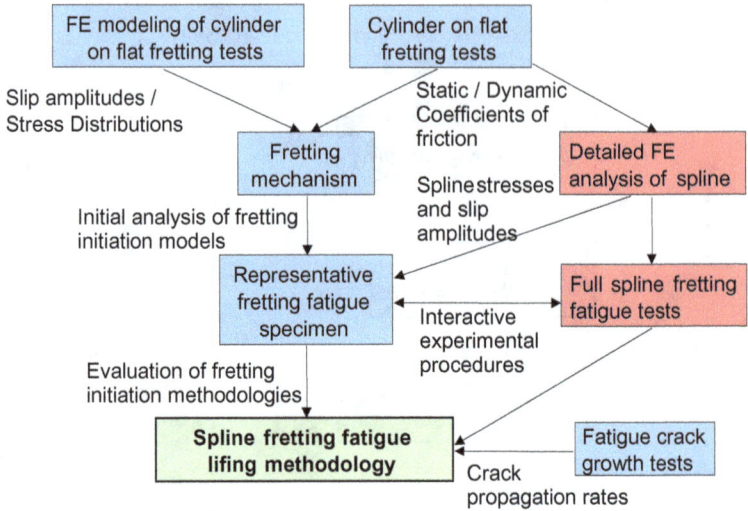

Fig. 23: Flowchart showing combined computational-experimental methodology proposed for development of URS test specimen and, hence, spline fretting fatigue lifing methodology [30].

Fig. 24: Photograph of simplified URS *in-situ* in servo-hydraulic fatigue test machine for fretting in spline couplings, showing strain gauges and LVDT for measuring force–displacement (hysteresis) responses and COF [31].

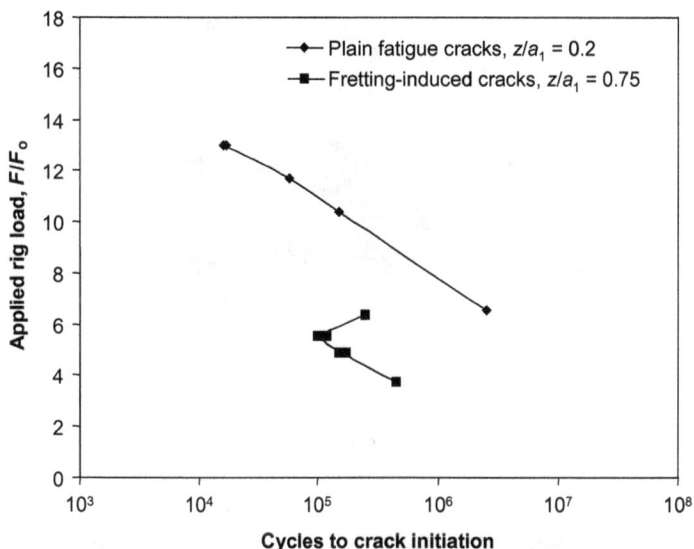

Fig. 25: Effect of applied load on URS plain fatigue ($z/a_1 = 0.2$) and fretting fatigue ($z/a_1 = 0.75$) behaviour. This suggests a critical slip range with a minimum fatigue life, similar to the "bath-tub" shape effect of Fig. 3 [31].

spline tooth interactions. The URS cyclic behaviour is consistent with that of the reduced scale spline coupling, with respect to both plain and fretting fatigue life for major cycle loading. The two axial locations on the reduced scale spline coupling simulated did not give fretting cracks within the 35,000 major load cycles (well beyond the design life) for either the spline coupling or URS under the design load case (Fig. 25). The URS was used to vindicate the use of nitriding of one contact surface, by demonstrating that, although nitrided specimens were more susceptible to fillet (e.g. spline root) cracking they were less susceptible to fretting (contact) cracking. The representative specimen approach offers a more economical and rapid turnover of test results, compared to spline coupling tests, and thus more rapid assessment of proposed designs, including geometry, materials, surface treatments, etc.

As the spline fatigue tests have indicated that fretting fatigue cracking is more readily caused by the combined effects of minor cycle rotating bending moment and fluctuating torque, superimposed on major cycle torque and axial load, a multiaxial RS (MRS) was subsequently developed to simulate the superposition of effects of minor and major cycle loads [32, 33], as shown in Fig 26. The MRS test rig mimics the history of combined torque, axial and bending load induced fretting variables within the

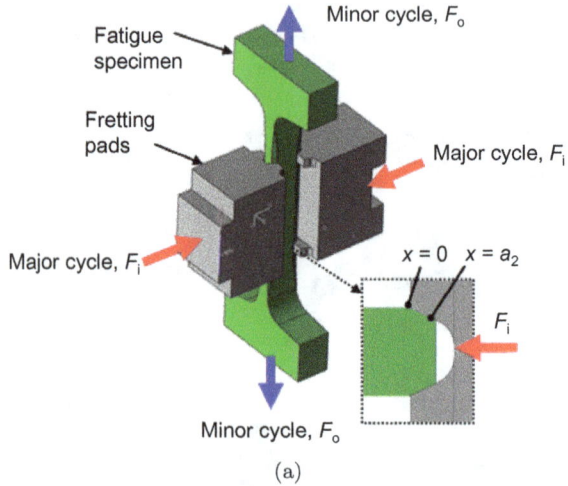

(a)

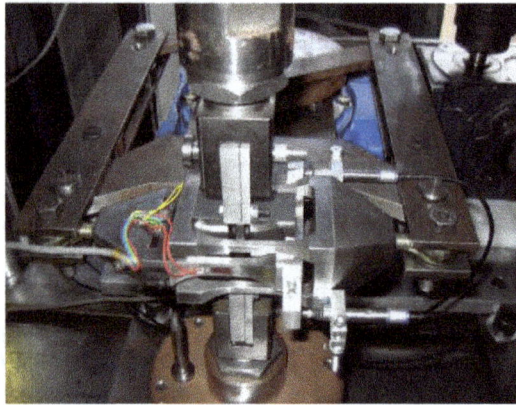

(b)

Fig. 26: (a) Schematic representation of the MRS, indicating the major and minor cycle loading and spline tooth geometry implemented in fretting bridge arrangement; (b) Photograph of *in-situ* MRS in servo-hydraulic fatigue test rig [32].

laboratory-scale coupling. The design of the simplified test is based on the FE-based identification of surface and sub-surface distributions of fatigue and fretting fatigue parameters at the experimental locations of fretting fatigue cracking in the scaled spline coupling tests (see Figs. 17–21). The multiaxial parameters chosen for approximate matching in the simplified test included (in-plane) contact pressure distribution, salient (out-of-plane) substrate fatigue stress and out-of-plane relative slip.

The key components are a fatigue specimen and a pair of fretting pads. The basic rationale is that the contact geometry of the pad-specimen interfaces (noting that there are four on each pad), are designed to generate a contact pressure distribution representative of that in spline teeth, under torque and axial loading, based on FE predictions for both. The fatigue specimen has chamfers, representative of the spline involute geometry, machined onto the outer surface along the gauge length. The pads have corresponding geometrical features so that the contact is representative of the spline tooth contact geometry. By adjusting the design variables of the fretting test geometry, specifically the fatigue specimen cross-sectional area and the fretting pad geometry (e.g. bridge stiffness, bridge gap, etc.) to achieve approximate convergence with respect to the key design variables mentioned above (including slip, contact pressure and substrate stresses), the final fretting test geometry is arrived at.

The fretting test is designed to operate in an existing electro servo-hydraulic machine. The fatigue specimen is a dog-bone type fatigue specimen, with the load transfer into the specimen effected via loading pins on which the shoulders of the specimen sit. The IP cyclic load source is a single phase electric motor controlled by a relay box and speed controller. The clamping device uses a cross-over clasp assembly to convert tensile forces from the motor-driven torque-arm assembly (driven by a 100:1 ratio worm drive gearbox) into compressive forces for clamping the fretting pads onto the fatigue specimen. The OP cyclic load source is provided by the servo-hydraulic ram and acts in the OP (vertical) direction.

Figure 27 shows a comparison of the stress-life results from the (i) plain fatigue tests on the spline high-strength CrMoV steel alloy, (ii) the laboratory-scale spline fretting fatigue tests, (iii) the simplified MRS tests and (iv) the spline FE model with wear-fatigue critical-plane SWT model. The stress used here for the spline results is the nominal maximum bending stress due to the rotating moment. This shows a dramatic knock-down (about 80%) in spline fatigue strength due to minor cycle induced fretting. However, the MRS test captures this effect and is consistently slightly conservative relative to the scaled spline tests. The FE model with wear effects also captures this effect, albeit is slightly non-conservative.

7. Simple Parameters for Prediction of Fretting

The competitive interaction between material removal and surface damage due to wear and fatigue crack nucleation due to surface damage is central to

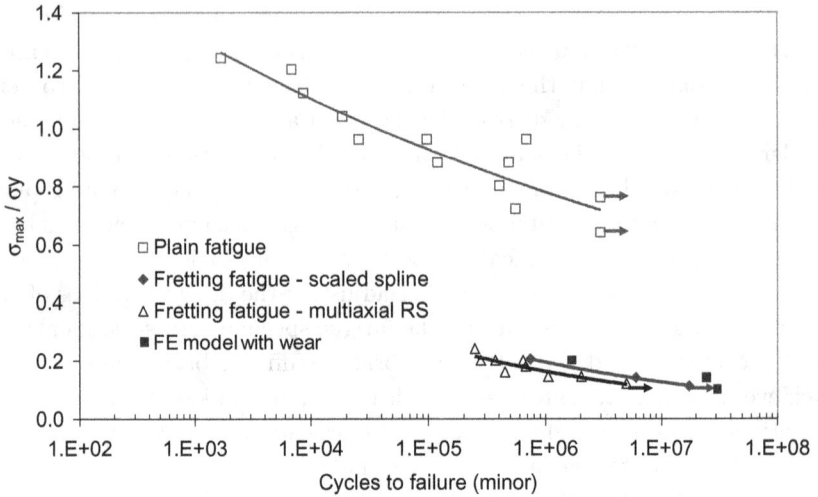

Fig. 27: Comparison of stress-life data from (i) plain fatigue tests on spline high-strength CrMoV steel alloy, (ii) laboratory-scale spline fretting fatigue tests, (iii) simplified multiaxial representative tests and (iv) spline FE model with wear-fatigue critical-plane SWT model.

fretting fatigue. From a modelling perspective, this was first demonstrated by Ding *et al.* [34] who showed how fretting wear simulation could allow prediction of the effect of sliding regime on multiaxial fatigue parameters, specifically (i) increasing stress–strain concentration at the stick–slip interface for partial slip with material removal due to wear cycles, as compared with (ii) decreasing trailing-edge tensile–shear stresses for gross-slip wear cycling, due to widening of the wear scar and associated re-distribution and attenuation of trailing-edge damaging stresses. Madge *et al.* [9] were the first to quantitatively predict the "bath-tub" shape effect of slip amplitude on fretting fatigue life (see Fig. 28) for fretting fatigue of Ti6Al4V. These effects are very important for predicting fretting fatigue crack initiation in complex couplings, such as splines. However, explicit simulation of material removal due to wear in FE models is computationally onerous, and prohibitively so for wear-fatigue cycle-by-cycle modelling of 3D couplings with multi-surface frictional contact, as shown earlier. Recently, Ding *et al.* [35] suggested a modified SWT approach for the prediction of fretting crack initiation, by introducing a fretting damage parameter D_{fret}. The introduction of D_{fret} is to emphasise the importance of frictional work $\tau\delta$ for crack nucleation under the fretting condition, especially in a small volume adjacent to the contact surface. D_{fret} aims to capture the additional influencing factors that are

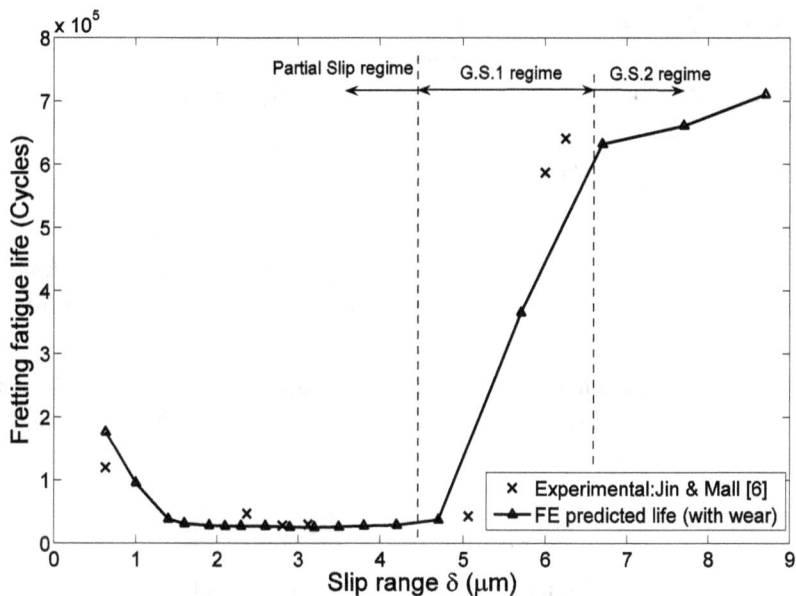

(a)

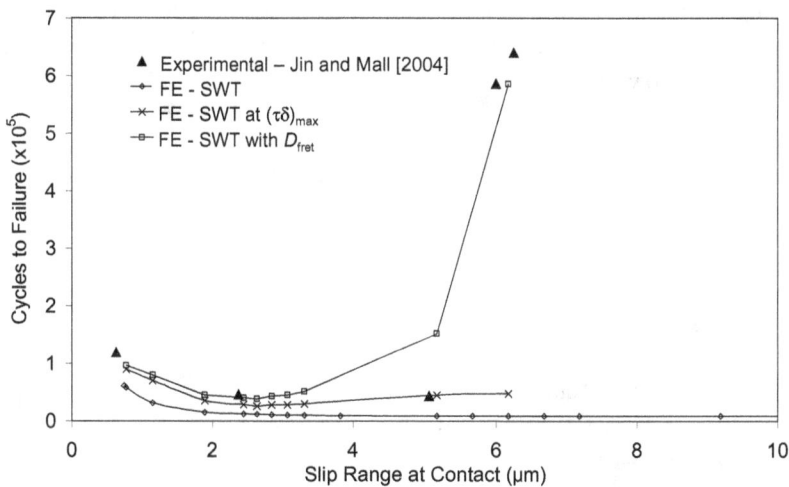

(b)

Fig. 28: (a) FE-predicted effect of slip amplitude on fretting fatigue life along with measured test data of Jin and Mall [36] for the same series of tests; (b) comparison of the predicted effect of slip amplitude on fretting fatigue life from D_{fret}.

neglected by general fatigue methods, and essentially comprises two competitive effects from frictional work $\tau\delta$: accelerating effect and retarding effect, as described earlier. The formulation of D_{fret} is given as

$$D_{\text{fret}} = (1 + C\tau\delta)\left\langle 1 - \frac{\tau\delta}{(\tau\delta)_{th}}\right\rangle^n, \qquad (11)$$

where the symbol $\langle\rangle$ is defined by $\langle u\rangle = u$ if $u > 0$ and $\langle u\rangle = 0$ if $u \leq 0$. $\tau\delta$ is the frictional work during one fretting cycle for a given local contact position. $(1 + C\tau\delta)$ is an empirical estimation of enhanced possibility of crack formation under the frictional work $\tau\delta$. $\left\langle 1 - \frac{\tau\delta}{(\tau\delta)_{th}}\right\rangle^n$ is introduced to characterise the effects of fretting wear. $(\tau\delta)_{th}$ is a threshold limit of $\tau\delta$ beyond which wear becomes dominant and there is no crack formation. The modified SWT approach is therefore expressed as

$$\sigma_{\max}\Delta\varepsilon_a D_{\text{fret}} = \frac{(\sigma_f')^2}{E}(2N_f)^{2b} + \sigma_f'\varepsilon_f'(2N_f)^{b+c} \quad \tau\delta \leq (\tau\delta)_{th}. \qquad (12)$$

The SWT-D_{fret} combined wear-fatigue parameter has been shown to capture (i) the "bath-tub" shape effect of relative slip for simple cylinder-on-flat fretting fatigue of Ti6Al4V, for example, as shown in Fig. 28(b), and (ii) the effect of minor cycle fretting on spline fatigue life and fretting crack location, as shown in Fig. 29.

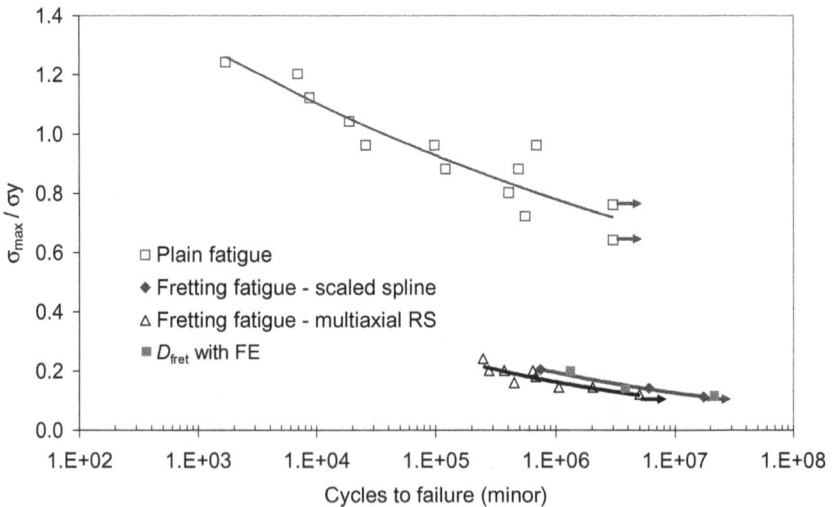

Fig. 29: Comparison of D_{fret} stress-life predictions against spline test data, MRS test data and plain fatigue data for spline steel.

8. Conclusions

Some of the key conclusions from this chapter are as follows:

1. A methodology has been demonstrated for finite element-based design of a multiaxial representative specimen test for fretting fatigue in helical, barrelled, aeroengine mainshaft spline couplings on the basis of matching of the fields of key fretting fatigue variables, specifically including contact tractions (pressure and shear), relative slip, subsurface stresses, as well as matching of contact surface (e.g. surface roughness, material, surface treatment) and representative local geometry.

2. The spline loading history, representative of in-service conditions, comprised combined torque, axial load and rotating bending moment, in a complex major (low frequency) and minor (higher frequency) cycle combination. In general, it was shown that in order to generate fretting fatigue cracks, significant overload (more than double the design loading conditions), in terms of minor-cycle rotating bending moment and fluctuating torque, superimposed on major cycle torque and axial load, was required. Major cycle loading alone was found to only lead to minor fretting wear.

3. Detailed, 3D finite element modelling of the barrelled, helical spline coupling, including frictional contact effects and multiaxial loading, has been key to the representative specimen test design.

4. The 3D, cycle-dependent mapping of contact-slip fields (slip, stick, open) for the effects of spline geometry (e.g. helical, involute, barrelling), deformation and friction evolution have permitted interpretation of the test results *vis-à-vis* the fretting regimes induced by the various effects of low-cycle, major cycle torque–axial loads (gross slip) and superimposed higher-cycle minor cycle rotating bending and fluctuating torque (mixed or partial slip).

5. A modified Archard equation wear approach, validated for cylinder-on-flat fretting, implemented for the spline coupling geometry, with barrelling and helical (involute) teeth, has been shown to be consistent with measured wear data from the spline testing.

6. The implementation of combined wear-fatigue modelling, on the basis of a critical-plane, multiaxial fatigue indicator parameter, previously validated for torque-axial spline loading fatigue, with Miner's rule for wear-induced multiaxial stress–strain evolution, has provided quantitative information for assessment of the effects of multiaxial fretting fatigue in splines. In particular, this includes prediction of major–minor cycles to cracking and fretting crack locations.

7. In order to circumvent the computationally intensive and expensive incremental wear-fatigue computer simulations of 3D, helical, barrelled (involute) spline couplings, a simplified fretting fatigue damage parameter, D_{fret}, for use with a multiaxial (critical-plane) fatigue indicator parameter (e.g. SWT) has been proposed. D_{fret} represents the effect of relative slip in terms of surface damage and has been shown to capture the well-known "bath-tub" shape for wear-fatigue interaction effect of slip amplitude in cylinder-on-flat laboratory fretting tests.

8. Careful finite element-based design of (i) a uniaxial representative specimen test gave results consistent with the spline major cycle loading in terms of plain and fretting fatigue and (ii) a multiaxial representative specimen test was shown to capture the effects of significant minor cycle overload-induced fretting fatigue, specifically reduction in fatigue strength.

Acknowledgements

The work on spline couplings described in this chapter represents the culmination of about ten years of collaboration and research at the University of Nottingham within the Rolls-Royce UTC in Gas Turbine Transmissions with significant contributions from many individuals, both researchers and technical staff, including Dr. Ian McColl (RIP), Prof. Tom Hyde, Dr. Ed Williams, Prof. Phil Shipway, Dr. Kenny Ding, Dr. Thomas R. Hyde, Dr. Jason Madge, Dr. Dean Houghton, Dr. Christian Ratsimba, Dr. Wei Siang Sum, Prof. Adib Becker, Dr. Paul Wavish, Dr. Sabesh Rajaratnam, Dr. Ian Richardson, Dr. Hianping Soh, Mr. Chun Ting Poon among others.

References

[1] R. B. Waterhouse, Fretting fatigue, *Int Mater Rev.* **37**(2), 77–97 (1992).

[2] O. J. McCarthy, J. P. McGarry, and S. B. Leen, Microstructure-sensitive prediction and experimental validation of fretting fatigue, *Wear.* **305**, 100–114 (2013).

[3] O. Vingsbo and D. Soderberg, On fretting maps, *Wear.* **126**, 131–147 (1988).

[4] S. B. Leen, T. R. Hyde, E. J. Williams, A. A. Becker, I. R. McColl, T. H. Hyde, and J. W. Taylor, Development of a representative test specimen for frictional contact in splined joint couplings, *J Strain Anal.* **35**(6), 521–544 (2000).

[5] S. B. Leen, I. J. Richardson, I. R. McColl, E. J. Williams, and T. R. Hyde, Macroscopic fretting variables in a splined coupling under combined torque and axial load, *J Strain Anal.* **36**(5), 481–499 (2001).

[6] J. Ding, I. R. McColl, and S. B. Leen, The application of fretting wear modelling to a spline coupling, *Wear.* **262**, 1205–1216 (2007).

[7] B. P. Volfson, Stress sources and critical stress combinations for splined shaft, *J Mech Des.* **104**, 551–556 (1982).

[8] J. Ding, W. Sum, S. Rajaratnam, S. B. Leen, I. R. McColl, and E. J. Williams, Fretting fatigue predictions in a complex coupling, *Int J Fatigue.* **29**, 1229–1244 (2007).

[9] J. J. Madge, I. R. McColl, and S. B. Leen, Contact-evolution based prediction of fretting fatigue life: effect of slip amplitude, *Wear.* **262**, 1159–1170 (2007).

[10] J. J. Madge, S. B. Leen, and P. H. Shipway, The critical role of fretting wear in the analysis of fretting fatigue, *Wear.* **263**, 542–551 (2007).

[11] Vincent, Y. Berthier, M. C. Dubourg, and M. Godet, Mechanics and materials in fretting, *Wear.* **153**, 135–148 (1992).

[12] K. L. Johnson, *Contact Mechanics.* Cambridge University Press, 1985.

[13] D. A. Hills and D. Nowell, *Mechanics of Fretting Fatigue.* Kluwer Academic Publishers, 1994.

[14] A. E. Giannakopoulos, T. C. Lindley, and S. Suresh, Aspects of equivalence between contact mechanics and fracture mechanics: theoretical considerations and a life-prediction methodology for fretting-fatigue, *Acta Metallurgica.* **46**(9), 2955–2968 (1998).

[15] A. Mugadu and D. A. Hills, A generalised stress intensity factor approach to characterising the process zone in complete fretting contacts, *Int J Solids Struct.* **39**, 1327–1335 (2002).

[16] M. Tur, J. Fuenmayor, A. Mugadu, and D. A. Hills, On the analysis of singular stress fields Part 1: finite element formulation and application to notches, *J Strain Anal.* **37**(5), 437–444 (2002).

[17] S. B. Leen, T. H. Hyde, C. H. H. Ratsimba, E. J. Williams, and I. R. McColl, An investigation of the fatigue and fretting performance of a representative aeroengine splined coupling, *J Strain Anal.* **37**(6), 565–583 (2002).

[18] C. H. H. Ratsimba, I. R. McColl, E. J. Williams, S. B. Leen, and H. Soh, Measurement, analysis and prediction of fretting wear damage in a representative aeroengine spline coupling, *Wear.* **257**, 1193–1206 (2004).

[19] W. Sum, S. B. Leen, E. J. Williams, S. Rajaratnam, and I. R. McColl, Efficient finite element modelling for complex shaft couplings under non-symmetric loading, *J Strain Anal.* **40**(7), 655–673 (2005).

[20] J. Ding, S. B. Leen, E. J. Williams, and P. H. Shipway, Finite element simulation of fretting wear-fatigue interaction in spline couplings, *Tribology.* **2**(1), 10–24 (2008).

[21] K. Ding, *Modelling of Fretting Wear,* PhD Thesis, University of Nottingham, UK, 2003.

[22] S. B. Leen, C. H. H. Ratsimba, I. R. McColl, and E. J. Williams, Fatigue life prediction for a barrelled splined coupling under torque overload, *Proc Instn Mech. Engrs, Part G: J Aerosp Eng.* **217**(G3), 123–142 (2003).

[23] I. R. McColl, J. Ding, and S. B. Leen, Finite element simulation and experimental validation of fretting wear, *Wear.* **256**, 1114–1127 (2004).

[24] S. Medina and A. V. Olver, An analysis of misaligned spline couplings, *Proc Instn Mech Engrs, Part J: J Eng Tribol.* **216**, 269–279 (2002).

[25] J. J. Madge, S. B. Leen, and P. H. Shipway, A combined wear and crack nucleation-propagation methodology for fretting fatigue prediction, *Int J Fatigue.* **30**(9), 1509–1528 (2008).

[26] M. A. Miner, Cumulative damage in fatigue, *J Appl Mech.* **12**, *Trans. ASME.* **67**, A159–A164 (1945).

[27] W. S. Sum, E. J. Williams, and S. B. Leen, Finite element, critical-plane, fatigue life prediction of simple and complex contact configurations, *Int J Fatigue.* **27**(4), 403–416 (2005).

[28] P. Lancaster and K. Salkauskas, *Curve and Surface Fitting: An Introduction.* Academic Press, 1986.

[29] D. Houghton, *Experimental and Computational Analysis of Multiaxial Fretting Fatigue in Spline Couplings,* PhD Thesis, University of Nottingham, UK, 2008.

[30] T. R. Hyde, *Development of a Representative Specimen for Fretting Fatigue of Spline Joint Couplings,* PhD Thesis, University of Nottingham, UK, 2002.

[31] T. R. Hyde, S. B. Leen, and I. R. McColl, A simplified fretting test methodology for complex shaft couplings, *Fat Fract Eng Mats Structs.* **28**(1), 10471067 (2005).

[32] P. M. Wavish, *Representative Specimen for Multiaxial Fretting Fatigue in a Splined Coupling,* PhD Thesis, University of Nottingham, UK, 2006.

[33] P. M. Wavish, D. Houghton, K. Ding, S. B. Leen, I. R. McColl, and E. J. Williams, A multiaxial fretting fatigue test for spline coupling contact, *Fat Fract Eng Mats Structs.* **32**, 325–345 (2009).

[34] J. Ding, S. B. Leen, and I. R. McColl, The effect of slip regime on fretting wear-induced stress evolution, *Int J Fatigue.* **26**, 521–531 (2004).

[35] J. Ding, D. Houghton, E. J. Williams, and S. B. Leen, Simple parameters to predict effect of surface damage on fretting fatigue, *Int J Fatigue.* **33**, 332–342 (2011).

[36] O. Jin and S. Mall, Effects of slip on fretting behaviour: experiments and analyses, *Wear.* **256**, 671–684 (2004).

Chapter 2

Fretting Fatigue Life Assessment by Accounting Wear

R. A. Cardoso[*], F. C. Castro[†], L. Reis[‡], and J. A. Araújo[†,§]

*Department of Mechanical Engineering, Federal University of Rio
Grande do Norte, 59078-970 Natal, RN, Brazil
†Department of Mechanical Engineering, University of Brasilia,
70910-900 Brasilia, DF, Brazil
‡IDMEC, Instituto Superior Técnico, Av. Rovisco Pais,
1049-001 Lisbon, Portugal
§alex07@unb.br

This chapter presents a numerical methodology for the fatigue assessment of mechanical components subjected to fretting fatigue. One knows that tribological interactions might significantly change fatigue damage mechanisms depending on the slip regime, environment conditions and materials investigated. Concerning the material loss caused by wear, near surface stresses might become milder or severer, which in turn will have a great impact in fatigue resistance. By focusing on the material loss due to wear in fretting contacts, this chapter presents a coupled numerical strategy, which accounts for both changes in the contacting surfaces geometry and the accumulation of fatigue damage by means of well-known multiaxial fatigue models. Throughout the text, theories presented are often confronted with experimental data. A discussion is also carried out concerning the applicability of neglecting wear effects when evaluating fretting fatigue under partial slip regime.

1. Introduction

Fretting fatigue involves mechanical assemblies subjected to both micro-slip between contacting surfaces and bulk fatigue loads acting in at least one of the mechanical parts. For some loading conditions, fretting loads may accelerate crack initiation and lead to premature failure. In spite of the great advances in the understanding of fretting fatigue over the last decades, which was accomplished by experimental, theoretical, and numerical studies, the phenomenon still raises many questions such as the competition between fatigue damage and wear processes. In the late 1980s, Vingsbo and Söderberg [1] published a well-known experimental work which suggested that wear might have a detrimental effect in fatigue life for partial slip conditions while it has a favourable effect for gross sliding regimes. The latter can be explained by the fact that, under gross sliding, the material loss due to wear becomes sufficiently high so that initiated cracks at the contact surface are worn out before they start propagating. Over the years, numerical studies have, in some way, confirmed these observations [2–6]. In this setting, it has been shown that, under partial slip conditions, the small amount of wear within the slip zones are responsible for slightly changing contacting geometries generating some sort of stress concentration, which increases fatigue damage processes and consequently crack initiation. On the other hand, during gross sliding, the higher amount of wear which takes place all over the contacting surfaces is responsible for increasing the contact region, which results in a reduction of contact tractions and, at the same time, the high wear rates constantly remove severely damaged areas, hindering the crack propagation process. Other works considering additional phenomena such as elastic–plastic behaviour [7, 8], variable friction coefficient [9]; and third body effect [10, 11] were also carried out, leading to similar conclusions. Despite the promising results found by the numerical modelling of fretting fatigue accounting wear, most of the life estimation approaches for fretting fatigue proposed in the literature neglect wear effects for partial slip conditions [12–15]. One of the main reasons for this is the high computational costs involved in the simulation of wear. Furthermore, due to the low wear rates under partial slip conditions, it is commonly assumed that the material loss can be neglected in the face of the high stress gradients beneath the contact. In this context, this chapter will focus on the development of a fatigue life estimation approach for fretting fatigue, which considers wear effects. Moreover, the implications of neglecting wear for life estimation of fretting fatigue problems will be examined.

2. Wear Modelling

To model wear, either a local version of Archard's law [2, 3] or the dissipated friction energy relation can be considered [6, 16]. Concerning the latter, its main advantage relies on the fact that it can easily account for changes in

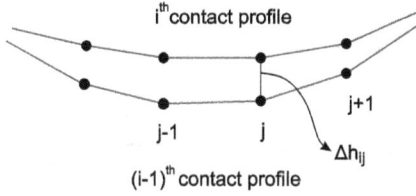

Fig. 1: Schematic representation of the contact nodes displacement representing wear material loss.

the friction coefficient between the contacting surfaces. In addition, both approaches can be readily implemented on a nodal basis in the finite element framework. When the dissipated energy relation is considered, the vertical displacement of the contact nodes representing the material loss due to wear can be obtained according to:

$$\Delta h_{ij} = \sum_{k=1}^{n_{\text{inc}}} \alpha q(x_j, t_k) \Delta s(x_j, t_k) \Delta N, \qquad (1)$$

where Δh_{ij} is the incremental wear depth for a given node j belonging to the contacting surface at the simulation of the ith fretting cycle (see Fig. 1), α is the dissipated energy wear coefficient, while $q(x_j, t_k)$ and $\Delta s(x_j, t_k)$ are the contact shear traction and the relative slip increment of the contact node located at the position x_j for the time instant t_k, respectively. The parameter ΔN in Eq. (1) is the jumping factor, which amplifies wear computations. This numerical strategy is used to speed up the simulation [2–6]; for instance, if one aims to perform wear computations for N_t fretting cycles, it is only necessary to simulate $N_t/\Delta N$ fretting cycles. In physical terms, the use of the jumping factor implies that wear is assumed to be constant for a given number of cycles ΔN. Note that the choice of this parameter has to be made judiciously to avoid instabilities in the solution of the problem [2]. In order to update the contacting surfaces during the simulation, either re-meshing [2, 3, 6, 17] or adaptive meshing techniques [4, 5, 17] can be considered. In the former, after simulating a complete fretting cycle, contact nodes are displaced by the amount specified in Eq. (1) and another mesh is generated in order to avoid element distortion, mainly under the gross sliding case. When the adaptive meshing strategy is used, position of contact nodes are also displaced according to Archard's law or Eq. (1), however, in this case, the mesh topology is held the same, whereas, for the nodes in the vicinity of the contacting surfaces, a mesh sweep process is carried out, which consists in shifting the position of such nodes in an

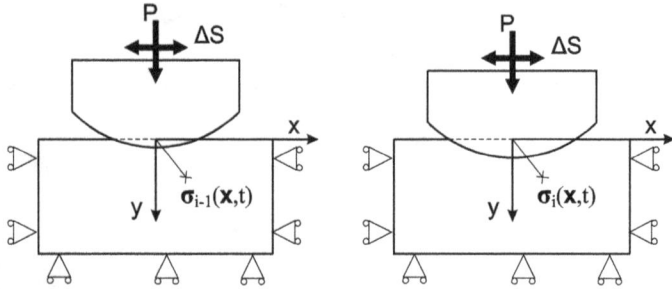

Fig. 2: Schematic view of the wear evolution between two consecutive simulated fretting cycles.

attempt to avoid mesh distortion. In this setting, an advection phase is further required to remap solution variables from the old mesh to the new one. It is worth mentioning that, when considering the adaptive meshing procedure, the position of the contact nodes might be performed incrementally within a fretting cycle, whereas for the re-meshing case, it is only possible after the simulation of an entire cycle.

Particularly, in Cardoso *et al.* [17] the re-meshing and adaptive meshing procedures were compared for both partial and gross slip regimes. In this case, authors concluded that the adaptive meshing strategy is computationally more efficient once it permits the use of higher cycle jumps in the wear modelling analyses. Besides, it does not require the unloading and re-loading phases of the normal force after the updating of the contact geometries as it is the case for the re-meshing procedure. However, the use of the adaptive meshing strategy is very memory-demanding, which might become a serious issue in the resolution of large problems. Furthermore, for partial slip conditions, the adaptive meshing technique struggles to provide a stabilised worn profile even for high enough number of fretting cycles [17].

When wear is included in the analysis, the subsurface stress field is constantly changing, Fig. 2. In this case, multiaxial fatigue parameters cannot be readily applied to the problem in an attempt to estimate fatigue life. A common way to overcome such an issue is to use an incremental damage formulation [4, 6], the simplest one being Miner's linear damage rule:

$$D_n = \sum_{i=1}^{n} \frac{\Delta N}{N_{f,i}}, \qquad (2)$$

where D_n is the damage accumulated up to the nth simulated fretting cycle (i.e. $n \times \Delta N$ physical wear cycles considered in the analysis) and

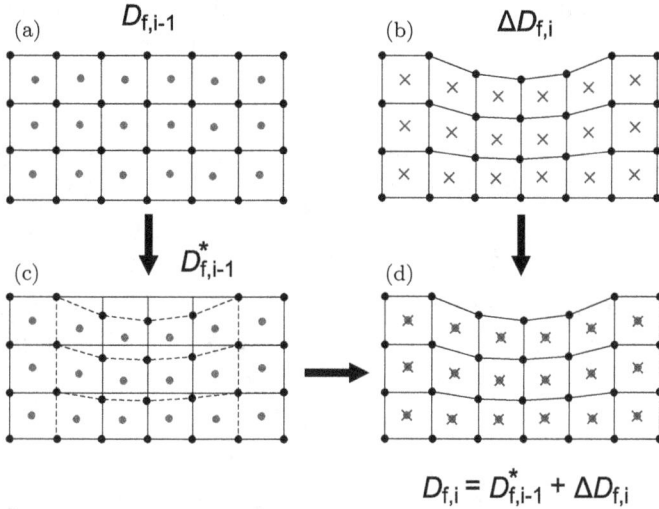

Fig. 3: Illustration of the methodology adopted to accumulate damage when considering wear: (a) accumulated damage at the centroid of the elements at the $(i-1)$th simulated fretting cycle, (b) incremental damage at the centroid of the elements at the ith simulated fretting cycles, (c) accumulated damage at the $(i-1)$th simulated fretting cycle extrapolated to the position of the element centroids at the ith simulated fretting cycle and (d) accumulation of the total damage at ith simulated fretting cycle [17].

$N_{f,i}$ is the fatigue life estimated for a given stress state at the ith simulated fretting cycle. Note that in the gross sliding regime the modification of the contacting surfaces due to wear may be significant. In this case, an interpolation/extrapolation process from one geometry configuration to another is necessary when making use of Eq. (2) [6, 17]. For instance, Fig. 3 illustrates a damage accumulation strategy for when the damage is defined at the centroid of the elements near the contacting surfaces [17].

3. Multiaxial Fatigue Models

In this section, a brief description of three well-known multiaxial fatigue models is given. The fatigue lives yielded by these models, for contact simulations neglecting and considering wear effects, will be evaluated in Section 5 using Ti-6Al-4V fretting fatigue data.

3.1. *Smith–Watson–Topper*

The Smith, Watson and Topper (SWT) criterion with a critical plane interpretation [18, 19] was developed for materials and loading conditions in which the crack initiation mechanism is predominantly governed by normal stresses and strains. The SWT model has the following form:

$$\text{SWT} = \sigma_{\text{nmax}}\varepsilon_{\text{na}}, \tag{3}$$

where σ_{nmax} and ε_{na} are the maximum normal stress and normal strain amplitude on the material plane that maximises their product.

3.2. Modified Wöhler curve method

The modified Wöhler curve method (MWCM) [20] is also used to estimate fatigue strength or life under complex loadings. The damage parameter used in the MWCM approach is expressed as

$$\text{MWCM} = \tau_{\text{a}} + \kappa \frac{\sigma_{\text{nmax}}}{\tau_{\text{a}}}, \tag{4}$$

where κ is a material constant and τ_{a} is the shear stress amplitude. In Eq. (4), the quantities σ_{nmax} and τ_{a} are evaluated on the critical plane, which is defined as the material plane that maximises τ_{a}. Inevitably, the same value of maximum τ_{a} appears in more than one material plane. In this case, the critical plane is chosen as the one where σ_{nmax} is higher. The MWCM approach is used for materials and loading conditions in which crack initiation is primarily governed by shear stresses.

3.3. Findley model

The critical plane model proposed by Findley [21] is expressed as

$$\text{FP} = (\tau_{\text{a}} + \gamma\sigma_{\text{nmax}})_{\text{max}}, \tag{5}$$

where γ is a material constant. In this model, the shear stresses are regarded as the primary cause of crack initiation and the normal stresses are responsible for increasing crack opening and thus the amount of fatigue damage. The critical plane is defined as the one that maximises the linear combination of τ_{a} and σ_{nmax}.

3.4. Definition of the shear stress amplitude

Multiaxial fatigue parameters generally require a definition for an equivalent shear stress amplitude, τ_{a}, since the shear stress vector, $\boldsymbol{\tau}$, on a material plane, Δ, may describe a generic path ψ over time, Fig. 4(a). Different methods for the calculation of τ_{a} have already been proposed [22, 23]. One of the most popular is the minimum circumscribed circle (MCC) [24]. However, this method presents a lack of sensitiveness with respect to the

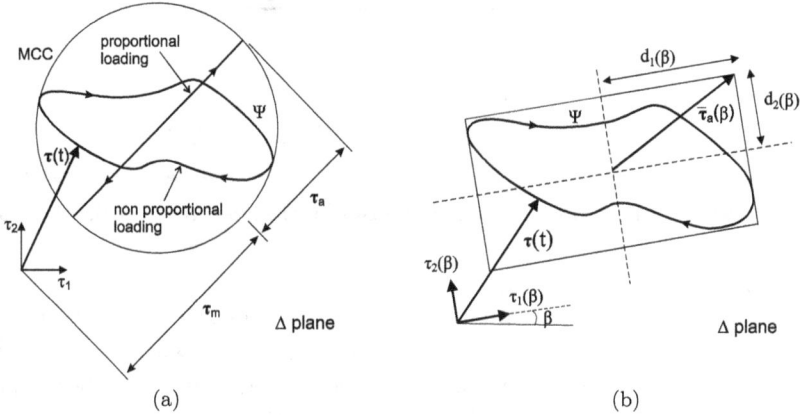

Fig. 4: Schematic representations of (a) the MCC and (b) the MRH for a given shear stress path ψ.

non-proportionality of the shear stress path, which may lead to inaccurate measurements of fatigue damage [25, 26]. An alternative method to compute τ_a capable of circumventing the aforementioned drawback was proposed by Mamiya *et al.* [27, 28]. These authors suggested that τ_a can be obtained by means of the maximum rectangular hull (MRH) enclosing the shear stress path ψ on a material plane Δ, Fig. 4(b). In this method, for each β-oriented rectangular hull, the shear stress amplitude is given by

$$\overline{\tau}_a(\beta) = \sqrt{d_1(\beta)^2 + d_2(\beta)^2} \tag{6}$$

and the equivalent shear stress amplitude, τ_a, is then defined by maximising $\overline{\tau}_a(\beta)$, i.e.

$$\tau_a = \max_{0° \leq \beta \leq 90°} \overline{\tau}_a(\beta). \tag{7}$$

4. Theory of Critical Distances

Fretting fatigue problems are inherently subjected to high stress gradients close to the contacting surfaces [29]. In this case, similarly to what happens in notch fatigue problems, hot-spot approaches are not suitable since they tend to underestimate the fatigue strength of the component [30–32]. In an attempt to develop a general framework to deal with fatigue problems involving stress gradients, the theory of critical distances (TCD) [33] first considered stress raisers such as notches and cracks, and then was successfully applied to fretting fatigue problems [34–36].

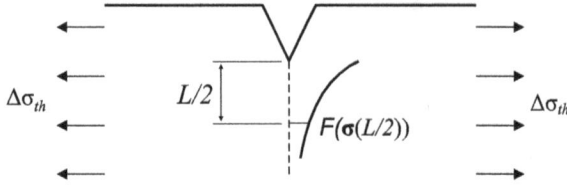

Fig. 5:   Notch under uniaxial loading where $\Delta\sigma_{\text{th}}$ is the stress range at the threshold condition.

The TCD consists in evaluating a fatigue criterion $F(\boldsymbol{\sigma})$ inside a process volume V surrounding a stress raiser and comparing it with λ, a constant that depends on the multiaxial fatigue criterion used. The size of the volume V is generally associated with El Haddad's length parameter given by [33]:

$$L = \frac{1}{\pi}\left(\frac{\Delta K_{\text{th}}}{\Delta\sigma_{-1}}\right)^2, \tag{8}$$

where ΔK_{th} is the threshold stress intensity factor range for long cracks and $\Delta\sigma_{-1}$ is the uniaxial fatigue limit range for fully reversed loading conditions. For 2D analysis, the TCD can be expressed in its simplified versions considering an area, a line or even a point [33]. Considering the Point Method, for instance, fatigue endurance is predicted to occur if

$$F(\boldsymbol{\sigma}(L/2)) \leq \lambda, \tag{9}$$

where the stress history is evaluated at a distance $L/2$ from the notch root (Fig. 5). Analogously, for fretting problems the stress history is computed at a distance $L/2$ beneath the contact edge (hot spot).

To extend the original TCD to fatigue life estimation, a relation between the critical distance and the number of cycles to failure has been proposed [14, 37, 38]. For example, under static failure the critical distance is given by [39]

$$L_s = \frac{1}{\pi}\left(\frac{K_{\text{Ic}}}{f\sigma_{\text{uts}}}\right)^2, \tag{10}$$

where K_{Ic} is the plane strain fracture toughness and f, a factor greater or equal to one that multiplies the ultimate tensile strength, σ_{uts}. A glance at Eqs. (8) and (10) suggests that L varies with the number of cycles to failure. In particular, the L vs. N relation can take the following power-law form [37, 40]:

$$L(N) = AN^B, \tag{11}$$

where A and B are constants that can be determined from fatigue life curves of both smooth and notched specimens [37, 40, 41] as illustrated in Fig. 6. In this case, just as an exemplification, the SWT model in conjunction with the TCD by means of the Volume Method is considered in the calibration procedure. As can be seen in Fig. 6, the calibration procedure uses as input the SWT vs. N_f relation determined from fatigue tests conducted on smooth specimens, as well as the S–N curve obtained by testing notched specimens. In this setting, for a given pair of observed fatigue life, N_{obs}, and fatigue stress amplitude, $\Delta S/2$, an FE simulation of the notched specimen reproducing the experimental configuration is conducted. In the following, by means of the Volume Method, the stress tensor, $\boldsymbol{\sigma}$, is averaged inside the fatigue process zone by considering a trial critical distance, L^{trial}. After that, such an average stress tensor works as input for the SWT fatigue model yielding the estimated fatigue life N_{SWT}, which in sequence has to be compared with the observed fatigue life, N_{obs}. If these two are close enough, the critical distance $L(N_{\mathrm{obs}})$ can be taken as L^{trial} and another fatigue data pair $(\Delta S/2, N_{\mathrm{obs}})$ might be assessed for the construction of the L vs. N relation. Otherwise, i.e. for $N_{\mathrm{SWT}} \neq N_{\mathrm{obs}}$, L^{trial} is modified and iterations are performed till N_{SWT} is sufficiently close to N_{obs}. After the evaluation of all the data in the S–N diagram, constants A and B of Eq. (11) are obtained by best fitting the generated L vs. N data points. Alternatively, in the lack of such fatigue data, A and B can be obtained by linearly interpolating Eqs. (8) and (10) for their corresponding number of cycles.

5. Wear Assessment for Partial Slip Fretting Fatigue

To investigate the implications of considering the material loss due to wear on life estimation under fretting fatigue conditions, experimental data for Ti-6Al-4V alloy were considered [17, 42]. The data were obtained by performing fretting fatigue tests [43] on the two-actuator fretting fatigue apparatus of the University of Brasilia, which is capable of independently controlling the bulk and fretting loads. The normal contact load is static and applied by a hydraulic system connected to an accumulator and a manual pump. A schematic view of the testing apparatus is shown in Fig. 7.

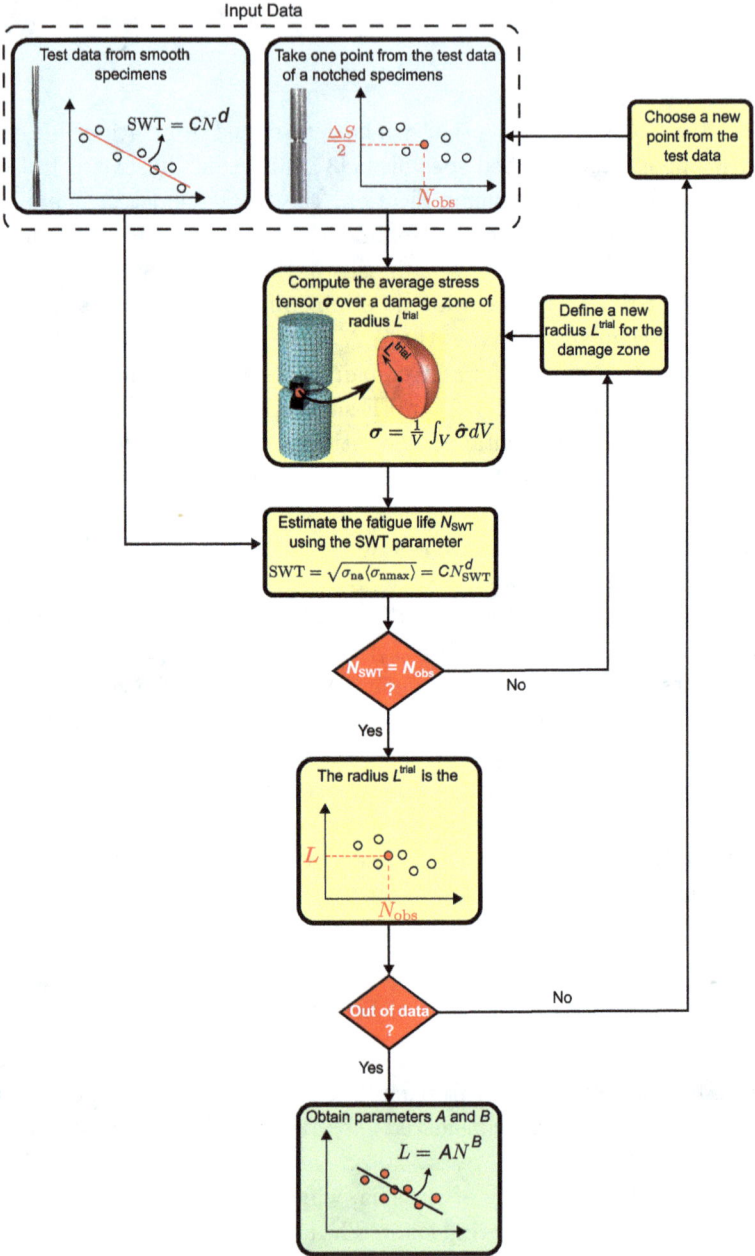

Fig. 6: Flowchart of the procedure to determine the L vs. N curve.

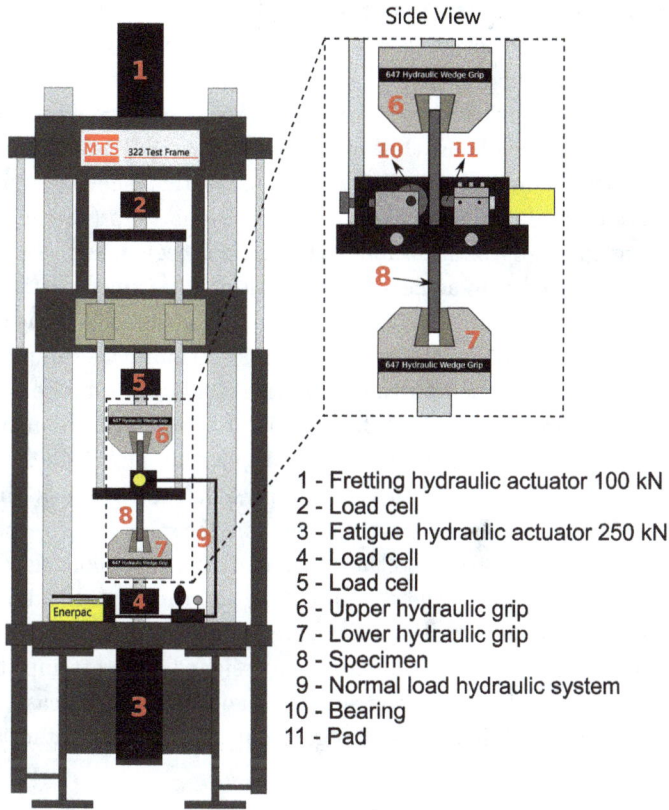

Fig. 7: Two-actuator fretting fatigue apparatus of the University of Brasilia.

The fretting fatigue apparatus, whose details of the contact system are shown in the side view of Fig. 7, is mounted on an MTS frame. The apparatus holds the pads carriage and the Enerpac static actuators (50 kN), which keeps the pressure on the pads. It can also operate using one pad and a bearing holder, as depicted in Fig. 7. Four metallic rods connect the pad carriage to the upper 100 kN hydraulic actuator (item 1) and its load cell (item 2). This actuator provides the tangential forces at the fretting contact. The 250 kN lower actuator (item 3) applies the bulk fatigue force to the specimen.

5.1. *Fretting fatigue data*

Cardoso *et al.* [17] and Araújo *et al.* [42] investigated the influence of considering wear on fretting fatigue life estimates when partial slip condition takes place. So, experimental data from [43] have been assessed. Such experimental data aimed to investigate size effects on fretting fatigue. Two sets of tests were designed and named Group-a and Group-b. In the Group-a tests, the aim was to investigate the influence on fatigue life of the volume of material being stressed beneath the contact surface. So, pairs of tests were designed experiencing partial slip conditions so that they had the same stress gradient close to the contact surface, the same slip (damaged) areas and different stressed volumes under the contact region, Fig. 8(a). Different stressed volumes were achieved by considering specimens with different thicknesses (8 and 13 mm). On the other hand, to ensure the same slip area for contacts with different thicknesses (8 and 13 mm), the amplitudes of the tangential load applied to the cylindrical pads were different. In this setting, the same stress gradient for a pair of tests could be attained by carefully choosing the bulk fatigue loads applied to each test. Note that, in Fig. 8, the parameters W, a and c represent the specimen thickness, contact semi-width and semi-width of the contact stick zone, respectively. The subscripts 8 and 13 refer to the specimen's thickness considered in the test. The theoretical size of the slip areas, A_{th}^s, on the other hand, is simply defined by the gray areas in the contact region.

Group-b tests, on the other hand, were designed in an attempt to verify the influence of the slip areas on fretting fatigue life. In this case, pairs of tests were designed so that the stress gradient close to the contact surfaces

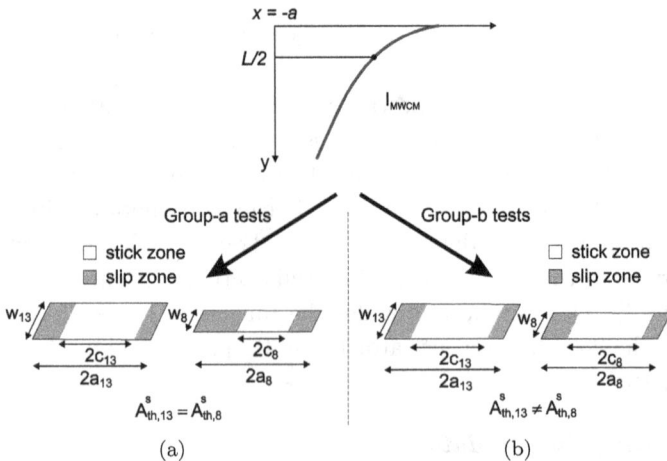

Fig. 8:   Schematic representation of the Group I tests: (a) Group-a tests with the same slip areas for different thicknesses, (b) Group-b tests with different slip areas [43].

were the same whereas the size of slip areas were different, Fig. 8(b). Such condition was easily obtained by only changing the specimens' thickness while keeping the same bulk load and contact loads per unit length. For both group of tests, the enforcement of the same stress gradient close to the contact edges for a given pair of tests were achieved by using the Point Method of the TCD [33] in association with the MWCM [20]. Details on the determination of the loading conditions and on the specifications of the pads and specimens can be found in Cardoso *et al.* [43].

In both group tests, the normal load, P, per unit length applied to the pads was the same, which resulted in contact zones with equal sizes $(2a)$ for all the tests. Partial slip conditions were ensured in all the tests. A schematic view of geometry and loading conditions used in the experiments is shown in Fig. 9. The loading history was prescribed as follows: first a mean bulk load F_{bm} is applied to the specimen on its right side. In sequence, a static normal load, P, presses the cylindrical pad against the flat specimen. After that, a sinusoidal tangential load Q is applied to the pad in phase with an alternate bulk load, F_{ba}, prescribed on the specimen as depicted in Fig. 9. The mean and alternate bulk loads originate the mean, σ_{bm}, and alternate, σ_{ba}, bulk stresses, respectively.

Table 1 summarises the loading conditions and observed fatigue life for the tests considered in Cardoso *et al.* (2019) and Araújo *et al.* (2020) while investigating the impact of considering wear on fretting fatigue life estimates for the partial slip regime. Their numerical investigation will be carefully addressed in Section 5.2. The variables σ_{bmax}, p_0, Q_{max}, A_s^{th} and N^{exp} in Table 1 refer to the maximum bulk load applied to the specimen, contact peak pressure (Hertz solution), the maximum tangential load per unit length, the theoretical size of the slip area and the observed fatigue life, respectively. Relevant mechanical properties of the Ti-6Al-4V are shown in Table 2. The friction coefficient, μ, adopted in the analyses was 0.5 [44].

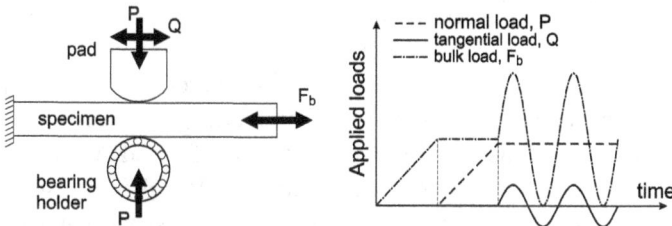

Fig. 9: Schematic view of the contact geometry and loading history in Cardoso *et al.* [43] fretting fatigue experiments.

Table 1:   Fretting fatigue data for the Ti-6Al-4V [17, 42].

Test	Thickness (mm)	σ_{bmax}/p_0	$Q_{max}/\mu P$	A_s^{th} (mm^2)	N^{exp}
T1	8	0.45	0.68	7.43	218×10^3
T2	13	0.60	0.46	7.37	167×10^3
T3	8	0.30	0.65	7.01	672×10^3
T4	13	0.46	0.44	6.99	443×10^3
T5	8	0.60	0.46	4.53	207×10^3
T6	8	0.40	0.42	4.08	529×10^3
T7	13	0.40	0.42	6.62	538×10^3

Table 2:   Ti-6Al-4V mechanical properties.

Material	E (GPa)	ν	σ_y (MPa)	σ_{uts} (MPa)	σ_{-1} (MPa)	ΔK_{th} (MPa·m$^{1/2}$)
Ti-6Al-4V	119.4	0.286	850	1000	583	5.5

Tests were carried out with a pad radius R of 70 mm, which produced a peak pressure p_0 of 500 MPa in all the tests.

5.2. *Assessment of fretting fatigue lives: Partial slip regime*

To assess the effects of including wear in partial slip fretting fatigue analyses, Cardoso [17] and Araújo *et al.* [42] compared life estimates taking wear into account with both life estimates neglecting wear and the experimental data presented in Section 5.1.

The wear modelling process was conducted based on an incremental FE framework. The vertical displacement of the contact nodes representing the material loss due to wear was obtained through Eq. (1) and contacting surfaces were updated by using a re-meshing technique. Once contacting surfaces as well as subsurface stresses constantly change due to wear, Miner's rule was applied to predict failure, Eq. (2) ($D = 1$). During the analysis, damage computations were performed at the centroid of the elements near the contact surface, Fig. 3. Figure 3 also describes the extrapolation procedure considered to accumulate Miner's damage parameter.

The FE model considered in Cardoso *et al.* [17] and Araújo *et al.* [42] (Fig. 10) was implemented into the FE software ABAQUS. A central refined zone surrounding the contacting surfaces, discretised with linear quadrilateral elements, was used in order to obtain accurate stress and strain solutions. The size of the mesh at this region was approximately 15 μm,

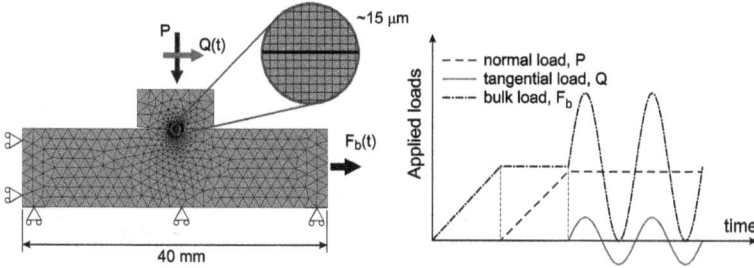

Fig. 10: FE model considered in the wear life estimation analysis [17, 42].

which yielded around 140 elements discretising the contact zone. Outside this region, triangular coarser elements were used. Plane strain elements were considered. Contact nodes on the pad were defined as the slave nodes while the ones in the specimen were defined as the master nodes. Frictional contact constraints were imposed via Lagrange multipliers. Rotation of the upper surface of the pad was prevented.

The use of Miner's rule, Eq. (2), demands the determination of the expected fatigue life, N_f, for material points given their stress–strain solutions over time. The latter are obtained during the fretting wear simulations, where N_f can be estimated by making use of any multiaxial fatigue criteria. In [17, 42], the multiaxial fatigue criteria assessed were the SWT, Findley's parameter and MWCM (Section 3). These models were calibrated based on uniaxial fatigue data for the Ti-6Al-4V [45], where the number of cycles to failure N_f could be related to each one of the aforementioned multiaxial fatigue parameters.

As fretting fatigue problems are commonly subjected to high stress gradients, the TCD has been used in the life estimates. In order to determine the critical distance curve presented in Eq. (11), data from static failure and fatigue limit conditions were used in [42] (see Section 4).

To estimate fretting fatigue life, two different approaches can be followed up. In one of them, wear effects such as the contact surface changes due to the material loss can be considered during the analysis. On the other hand, wear implications can be entirely neglected by assuming that wear rates are very low and do not play a significant role in the fatigue damage processes, which is actually assumed in many works in the field [12–15, 46]. In case of disregarding the effects of wear, life can be directly estimated by evaluating any multiaxial fatigue criterion at the centre of the fatigue process zone (Point Method), which is $L/2$ vertically distant from the contact trailing edge (hot spot), Fig. 11. Note that cracks are not necessarily

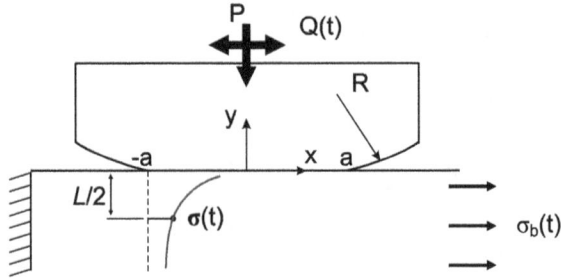

Fig. 11:   TCD applied to fretting fatigue problems when wear is neglected.

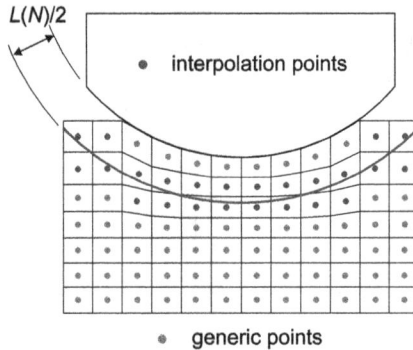

Fig. 12:   TCD applied to fretting fatigue problems when considering wear [42].

initiated perpendicular to the contact surface as might be suggested in Fig. 11. Such representation is the standard approach usually considered in the implementation of the Point Method in spite of the crack initiation mechanisms. Besides, although the Point Method has been considered in [17, 42], other averaging approaches such as a line, an area or even a volume (3D problems) could also be utilised. On the contrary, if wear effects are included in the analysis, Miner's rule needs to be considered once subsurface stress fields constantly change. Furthermore, the hot-spot point at the contact surface is not fixed over the fretting cycles either, due to changes in the contacting surfaces. In this case, one possible strategy to be adopted is to cumulate increments of Miner's damage parameter but only considering points $L/2$ vertically distant from the contact surface [42], Fig. 12. Note that, by either considering or disregarding wear effects, the TCD is used once fretting problems inherently result in high stress gradients.

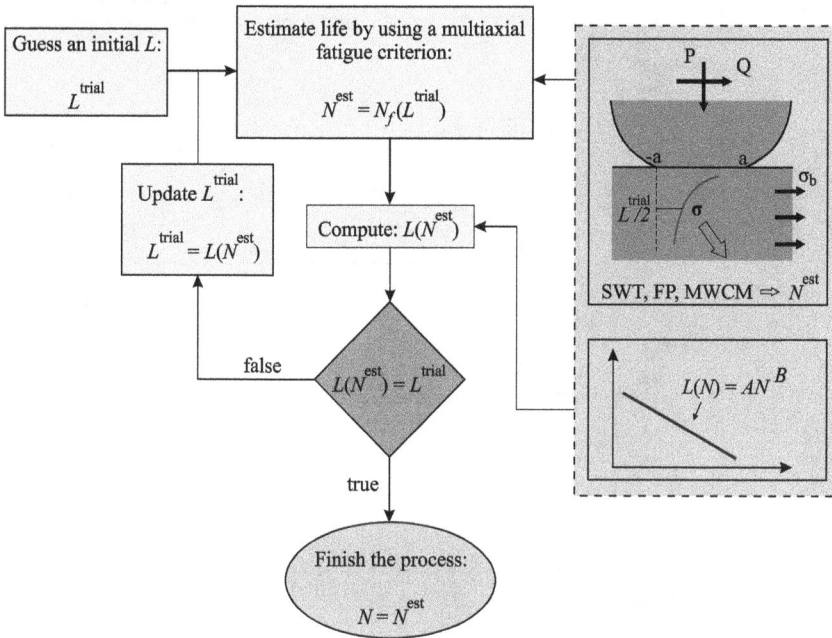

Fig. 13:   Flowchart of the life estimation methodology when neglecting wear.

When neglecting wear, life can be estimated through an iterative approach, which is summarised in Fig. 13. In this case, first an initial guess for the critical distance is assumed (L^{trial}). Then, the estimated life, N^{est}, can be obtained by evaluating a given multiaxial fatigue criterion at $L^{\text{trial}}/2$ (Fig. 11). In the following, N^{est} is used to compute a new critical distance $L(N^{\text{est}})$ through Eq. (11). If the latter is close enough to L^{trial}, the process is finished and the estimated life $N = N^{\text{est}}$ is obtained. Otherwise, L^{trial} is replaced by $L(N^{\text{est}})$ and the iterative process continues until $L(N^{\text{est}})$ is close enough to L^{trial}.

On the other hand, if the material loss due to wear is considered, instead of using an iterative approach, an incremental strategy has to be adopted, Fig. 14. In this case, we start considering the initial contact geometries in order to perform the simulation of an initial fretting cycle. Assuming that the contacting surfaces remain unchanged for ΔN fretting cycles, the damage (D_1) generated for such number of cycles ($N_1 = \Delta N$) can be computed by using Eq. (2) for each element centroid close to the contacting surface (Fig. 12). Note that element gauss points or even nodal points can also be considered at this stage. In the following, the critical distance parameter is

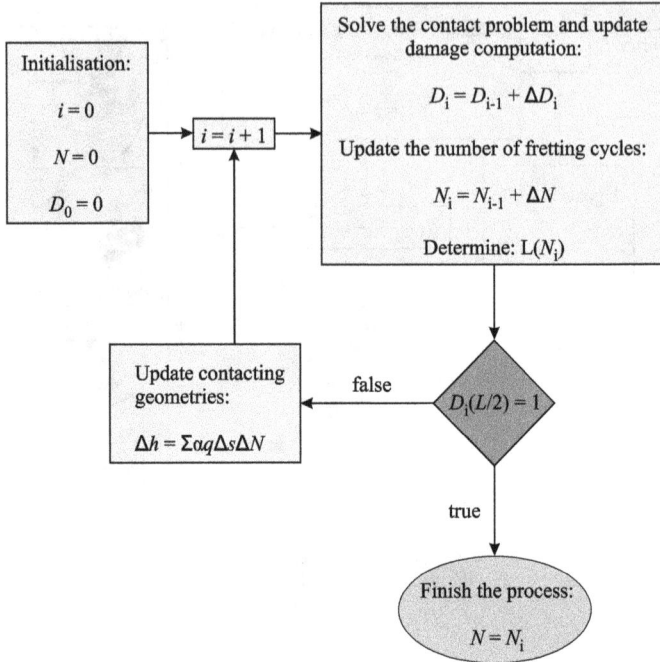

Fig. 14:   Flowchart of the life estimation strategy when considering wear.

computed considering the given number of fretting cycles N_1. Then, damage computations lying on a line $L(N_1)/2$ beneath the contact surface are verified for fatigue failure, i.e. $D_1 \geq 1$. If so, the processes are terminated and the predicted life is N_1. Otherwise, contact geometries are updated with the help of Eq. (1) and the process continues, i.e. a new fretting cycle is simulated by assuming that such contact configuration remains the same for another ΔN cycles. In the following, damage D_1 for points near the contacting surfaces must be increased by the damage increment ΔD_2 generated by the new ΔN cycles, which results in the cumulated damage D_2 for the first $N_2 = N_1 + \Delta N$ fretting cycles. Next, this damage is checked for failure, i.e. if $D_2 \geq 1$ for any point lying on the curve $L(N_2)/2$ beneath the contact surface, the process is finished and the estimated fatigue life is given by N_2. Otherwise, contact geometries are updated and the previous steps are repeated for i steps until $D_i \geq 1$ holds true.

Assuming that the jumping factor ΔN is small enough, D_i will be close to one and the estimated life can be taken as N_i (i.e. $i \times \Delta N$).

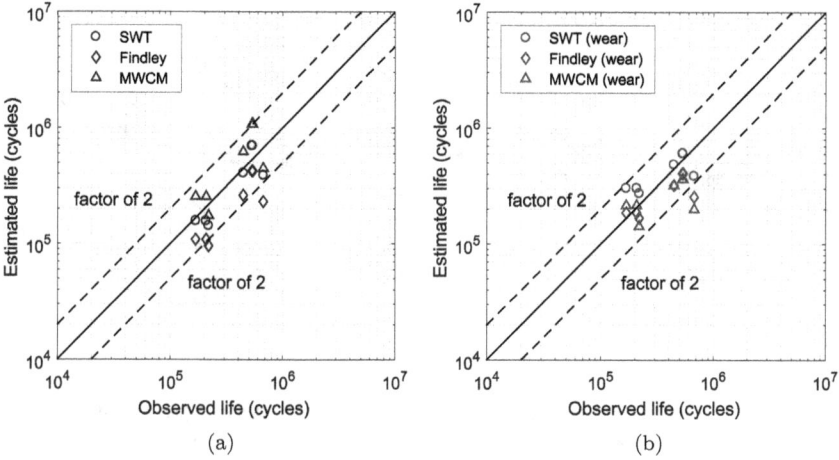

Fig. 15: Estimated vs. observed fretting fatigue lives: (a) neglecting and (b) considering wear in the analyses [42].

Damage assessments for a line $L/2$ beneath the contact surface are obtained by interpolating damage data in the surroundings of such a line, Fig. 12.

After applying the aforementioned life estimate strategies to the experimental data described in Section 5.1, the influence of considering wear effects in fretting fatigue analysis was considered. Figure 15 shows life estimates neglecting (a) and including (b) wear effects. Circle markers represent results obtained when the SWT fatigue parameter is used. Diamond and triangle markers refer to life estimates considering Findley's parameter and the MWCM, respectively. It is seen that both approaches provide satisfactory results as most of the life estimates were within the factor-of-two boundaries. However, it is also seen that the results accounting wear are slightly more centered (Fig. 15(b)).

The present analysis indicates that conventional life estimation approaches, which usually neglect wear, can provide satisfactory results. Furthermore, the slight improvement obtained when wear is considered might not justify the computational cost added to the simulations. Araújo *et al.* [42] also compared life estimates obtained by considering a variable critical distance (Fig. 15) with those assuming a constant L given by Eq. (8), see Fig. 16. When comparing Figs. 15 and 16, one can see that the use of a variable critical distance provided somewhat better results.

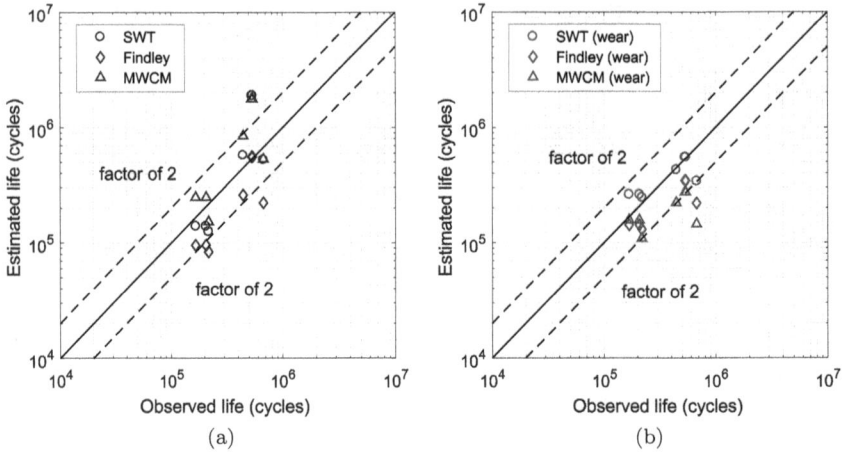

Fig. 16: Estimated vs. observed fretting fatigue by assuming a constant L: (a) results of neglecting and (b) accounting wear in the analyses [42].

6. Assessment of Fretting Fatigue Lives — Gross Sliding Regime

Different to what has been concluded by Araújo *et al.* [42] for partial slip regime, where life estimates neglecting wear provided satisfactory results, the numerical work of Pinto *et al.* [47] has shown that, for gross sliding conditions, the effects of wear cannot be disregarded. In this case, the high wear rates associated with the reduction of contact stresses over cycles have beneficial effects in terms of fatigue life. For example, Fig. 17 shows the maximum damage (D_{max}) evaluated over a line $L/2$ beneath the contact surface for a fretting fatigue configuration leading to gross sliding regime. As can be seen, damage computations are compared for when wear is taken into account (solid line) and neglected (dashed line). For a given number of cycles, when accounting wear, the accumulated damage finds a plateau differently to what is observed when disregarding the material loss due to wear, where material failure would be prematurely predicted. Similar trends were already observed in other works [4, 6]. Therefore, it must be highlighted that, even though wear effects might be neglected in fretting fatigue when partial slip conditions prevail, the same is not true for gross sliding situations and life estimation strategies like the one presented in Section 5.2 have to be considered.

Another result that illustrates the impact of wear in fatigue life assessment when gross sliding takes place can also be found in the work of Pinto

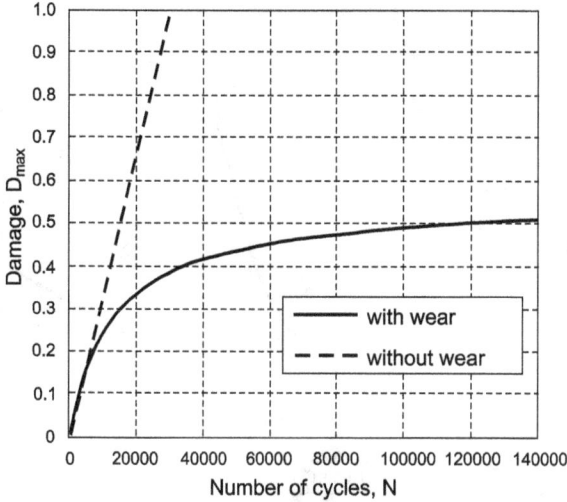

Fig. 17: Damage computations for an aluminium alloy under gross slip fretting fatigue conditions by considering and neglecting wear [47].

et al. [47], which numerically investigated fretting fatigue under variable amplitude loading. In this case, one of the loading sequences investigated in such work consisted in prescribing a given number of fretting cycles, n_1, under partial slip conditions followed by the application of n_2 loading cycles experiencing gross sliding conditions (L–H loading sequence). The opposite was also investigated, i.e. a sequence of n_1 gross sliding loading cycles followed by the application of n_2 partial slip fretting cycles (H–L loading sequence). In all the cases, the fatigue bulk and contact normal loads were the same while the tangential load/displacement was changed between each slip regime. For each one of these loading sequences, the fatigue damage generated by the first loading block (n_1 cycles) was limited to a given value d_1. The second block was then prescribed over n_2 cycles generating the fatigue damage d_2. Fatigue failure was predicted whenever $d_1 + d_2 = 1$. Fatigue damage was evaluated at a constant critical distance.

Results in Fig. 18 depict the life estimation results for the H–L and L–H loading sequences in terms of the damage d_1 generated by the first n_1 fretting cycles belonging to the first loading sequence. As can be seen, for the H–L loading sequence, finite life is only observed if d_1 is less than a certain value. It happens once for high number of fretting cycles under gross sliding conditions (High block), the material removal becomes severe, which increases contact size and reduces contact stresses. Such contact

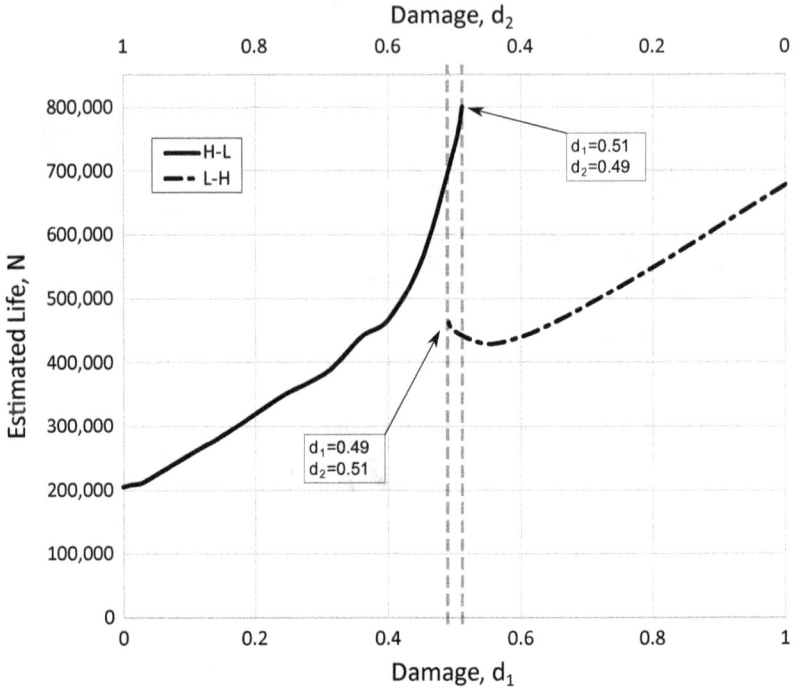

Fig. 18:   Estimated life for the variable amplitude loading sequences H–L and L–H in terms of the fatigue damage in the first loading block [47].

stress reduction hinders fatigue failure in the subsequent partial slip loading sequence (Low block). On the other hand, for the L–H loading sequence, finite life is only observed when d_1 is greater than a certain value. In this case, unless the damage generated under partial slip regime is high enough, the high wear rate in the subsequent gross sliding loading sequence (High block) will remove severely damaged areas resulting in infinite life. These results once more highlight the impact of wear in the fretting fatigue analysis of problems experiencing gross sliding. It is worth noting that, as the aforementioned analyses were based on numerical simulations, tests should be conducted under gross sliding regime to validate these results.

7. Final Remarks

This chapter discussed the implications of including wear into a predictive fretting fatigue life methodology. The ingredients of such a methodology relied on the application of well-established multiaxial fatigue life criteria in conjunction with the Theory of Critical Distances and Miner's linear damage rule. Simulation of the material loss due to wear was addressed by using a local dissipated friction energy rule, which can be easily implemented on a nodal basis in finite element analysis. The estimated lives were compared with experimental fretting fatigue data obtained under partial slip conditions. Results have shown that the simplified approach in which wear effects' affects are disregarded provides most of the life estimates within a factor of 2 band. It was also shown that considering a variable critical distance can improve accuracy of life estimates.

Concerning the gross sliding condition, our results indicated that wear effects should not be neglected. The high material loss in the gross sliding regime leads to less severe fatigue damage due to the constant removal of surface layers and contact tractions redistribution.

References

[1] O. Vingsbo and S. Söderberg, On fretting maps, *Wear.* **126**(2), 131–147 (1988).

[2] I. R. McColl, J., Ding, and S. B. Leen, Finite element simulation and experimental validation of fretting wear, *Wear.* **256**(11–12), 1114–1127 (2004).

[3] J. Ding, S. B., Leen, and I. R. McColl, The effect of slip regime on fretting wear-induced stress evolution, *Int J Fatigue.* **26**(5), 521–531 (2004).

[4] J. J. Madge, S. B. Leen, I. R. McColl, and P. H. Shipway, Contact-evolution based prediction of fretting fatigue life: Effect of slip amplitude, *Wear.* **262**(9–10), 1159–1170 (2007).

[5] A. Cruzado, S. B. Leen, M. A. Urchegui, and X. Gómez, Finite element simulation of fretting wear and fatigue in thin steel wires, *Int J Fatigue.* **55**, 7–21 (2013).

[6] S. Garcin, S. Fouvry, and S. Heredia, A fem fretting map modeling: Effect of surface wear on crack nucleation, *Wear.* **330**, 145–159 (2015).

[7] A. L. M. Tobi, J. Ding, G. Bandak, S. B. Leen, and P. H. Shipway, A study on the interaction between fretting wear and cyclic plasticity for ti–6al–4v, *Wear.* **267**(1–4), 270–282 (2009).

[8] F. Shen, W. Hu, and Q. Meng, A damage mechanics approach to fretting fatigue life prediction with consideration of elastic–plastic damage model and wear, *Tribol Int.* **82**, 176–190 (2015).

[9] T. Yue and M. A. Wahab, Finite element analysis of fretting wear under variable coefficient of friction and different contact regimes, *Tribol Int.* **107**, 274–282 (2017).

[10] J. Ding, I. R. McColl, S. B. Leen, and P. H. Shipway, A finite element based approach to simulating the effects of debris on fretting wear, *Wear.* **263**(1–6), 481–491 (2007).

[11] S. Basseville, E. Héripré, and G. Cailletaud, Numerical simulation of the third body in fretting problems, *Wear.* **270**(11–12), 876–887 (2011).

[12] T. Hattori, M. Nakamura, and T. Watanabe, Simulation of fretting-fatigue life by using stress-singularity parameters and fracture mechanics, *Tribol Int.* **36**(2), 87–97 (2003).

[13] C. Navarro, S. Muñoz, and J. Domínguez, On the use of multiaxial fatigue criteria for fretting fatigue life assessment, *Int J Fatigue.* **30**(1), 32–44 (2008).

[14] J. A. Araújo, L. Susmel, M. S. T. Pires, and F. C. Castro, A multiaxial stress-based critical distance methodology to estimate fretting fatigue life, *Tribol Int.* **108**, 2–6 (2017).

[15] C. F. B. Sandoval, L., Malcher, F. A. Canut, L. M. Araújo, T. C. R. Doca, and J. A. Araújo, Micromechanical gurson-based continuum damage under the context of fretting fatigue: Influence of the plastic strain field, *Int J Plasticity.* **125**, 235–264 (2020).

[16] S. Fouvry, T. Liskiewicz, P. Kapsa, S. Hannel, and E. Sauger, An energy description of wear mechanisms and its applications to oscillating sliding contacts, *Wear.* **255**(1–6), 287–298 (2003).

[17] R. A. Cardoso, T. Doca, D. Néron, S. Pommier, and J. A. Araújo, Wear numerical assessment for partial slip fretting fatigue conditions, *Tribol Int.* **136**, 508–523 (2019).

[18] R. N. Smith, P. Watson, and T. H. Topper, A stress-strain parameter for the fatigue of metals, *J Mater.* **5**(4), 767–778 (1970).

[19] D. F. Socie, Multiaxial fatigue damage models, *J Eng Mater Technol.* **109**(4), 293–298 (1987).

[20] L. Susmel and P. Lazzarin, A bi-parametric wöhler curve for high cycle multiaxial fatigue assessment, *Fatigue Fract Eng M.* **25**(1), 63–78 (2002).

[21] W. N. Findley, A theory for the effect of mean stress on fatigue of metals under combined torsion and axial load or bending, *J Eng Ind.* **81**(4), 301–305 (1959).

[22] L. Susmel, A simple and efficient numerical algorithm to determine the orientation of the critical plane in multiaxial fatigue problems, *Int J Fatigue.* **32**(11), 1875–1883 (2010).

[23] F. C. Castro, J. A. Araújo, E. N. Mamiya, and P. A. Pinheiro, Combined resolved shear stresses as an alternative to enclosing geometrical objects as a measure of shear stress amplitude in critical plane approaches, *Int J Fatigue.* **66**, 161–167 (2014).

[24] V. K. Dang, Sur la résistance à la fatigue des métaux, *Sci Tech Armement.* **47**, 429–453 (1973).

[25] I. V. Papadopoulos, A new criterion of fatigue strength for out-of-phase bending and torsion of hard metals, *Int J Fatigue.* **16**(6), 377–384 (1994).

[26] B. Li, L. Reis, and M. De Freitas, Comparative study of multiaxial fatigue damage models for ductile structural steels and brittle materials, *Int J Fatigue.* **31**(11–12), 1895–1906 (2009).

[27] E. N. Mamiya, J. A., Araújo, and F. C. Castro, Prismatic hull: A new measure of shear stress amplitude in multiaxial high cycle fatigue, *Int J Fatigue.* **31**(7), 1144–1153 (2009).

[28] J. A. Araújo, A. P. Dantas, F. C. Castro, E. N. Mamiya, and J. L. A. Ferreira, On the characterization of the critical plane with a simple and fast alternative

measure of the shear stress amplitude in multiaxial fatigue, *Int J Fatigue.* **33**(8), 1092–1100 (2011).

[29] J. A. Araujo and D. Nowell, The effect of rapidly varying contact stress fields on fretting fatigue, *Int J Fatigue.* **24**(7), 763–775 (2002).

[30] N. E. Frost and D. S. Dugdale, Fatigue tests on notched mild steel plates with measurements of fatigue cracks, *J Mech Phys Solids.* **5**(3), 182–192 (1957).

[31] G. Meneghetti, L. Susmel, and R. Tovo, High-cycle fatigue crack paths in specimens having different stress concentration features, *Eng Fail Anal.* **14**(4), 656–672 (2007).

[32] F. C. Castro, J. A. Araújo, and N. Zouain, On the application of multiaxial high-cycle fatigue criteria using the theory of critical distances, *Eng Fract Mech.* **76**(4), 512–524 (2009).

[33] D. Taylor, Geometrical effects in fatigue: A unifying theoretical model, *Int J Fatigue* **21**(5), 413–420 (1999).

[34] J. A. Araújo, L. Susmel, D. Taylor, J. C. T. Ferro, and E. N. Mamiya, On the use of the theory of critical distances and the modified wöhler curve method to estimate fretting fatigue strength of cylindrical contacts, *Int J Fatigue.* **29**(1), 95–107 (2007).

[35] J. A. Araújo, L. Susmel, D. Taylor, J. C. T. Ferro, and J. L. A. Ferreira, On the prediction of high-cycle fretting fatigue strength: Theory of critical distances vs. hot-spot approach, *Eng Fract Mech.* **75**(7), 1763–1778 (2008).

[36] S. Fouvry, H. Gallien, and B. Berthel, From uni-to multi-axial fretting-fatigue crack nucleation: Development of a stress-gradient-dependent critical distance approach, *Int J Fatigue.* **62**, 194–209 (2014).

[37] L. Susmel and D. Taylor, A novel formulation of the theory of critical distances to estimate lifetime of notched components in the medium-cycle fatigue regime, *Fatigue Fract Eng M.* **30**(7), 567–581 (2007).

[38] C. T. Kouanga, J. D. Jones, I. Revill, A. Wormald, D. Nowell, R. S. Dwyer-Joyce, J. A. Araújo, and L. Susmel, On the estimation of finite lifetime under fretting fatigue loading, *Int J Fatigue.* **112**, 138–152 (2018).

[39] L. Susmel and D. Taylor, On the use of the theory of critical distances to predict static failures in ductile metallic materials containing different geometrical features, *Eng Fract Mech.* **75**(15), 4410–4421 (2008).

[40] L. Susmel and D. Taylor, The modified wöhler curve method applied along with the theory of critical distances to estimate finite life of notched components subjected to complex multiaxial loading paths, *Fatigue Fract Eng M.* **31**(12), 1047–1064 (2008).

[41] P. Rocha, J. Diaz, C. Silva, J. Araújo, and F. Castro, Fatigue of two contacting wires of the acsr ibis 397.5 mcm conductor: Experiments and life prediction, *Int J Fatigue.* **127**, 25–35 (2019).

[42] J. A. Araújo, F. C. Castro, I. M., Matos, and R. A. Cardoso, Life prediction in multiaxial high cycle fretting fatigue, *Int J Fatigue.* **134**, 105504 (2020).

[43] R. A. Cardoso, B. Ferry, C. Montebello, J. Meriaux, S. Pommier, and J. A. Araújo, Study of size effects in fretting fatigue, *Tribol Int.* **143**, 106087 (2020).

[44] J. Bellecave, S. Pommier, Y. Nadot, J. Meriaux, and J. A. Araújo, T-stress based short crack growth model for fretting fatigue, *Tribol Int.* **76**, 23–34 (2014).

[45] A. R. Kallmeyer, A. Krgo, and P. Kurath, Evaluation of multiaxial fatigue life prediction methodologies for ti-6al-4v, *J Eng Mater Technol.* **124**(2), 229–237 (2002).

[46] R. A. Cardoso, E. R. F. d. S. Campos, J. L. A. Ferreira, D. Wang, and J. A. Araújo, A crack arrest methodology based on bazants parameter to fretting fatigue, *Theor Appl Fract Mec.* **95**, 208–217 (2018).

[47] A. L. Pinto, R. A. Cardoso, R. Talemi, and J. A. Araújo, Fretting fatigue under variable amplitude loading considering partial and gross slip regimes: Numerical analysis, *Tribol Int.* **146**, 106199 (2020).

Chapter 3

Numerical Frameworks for the Wear Modelling of Key Engineering Materials

T. Doca

Department of Mechanical Engineering, University of Brasilia,
70910-900, Brasilia, Brazil
doca@unb.br

A comparison of two numerical frameworks designed for wear analysis is presented. Experimental data regarding three different wear regimes (abrasion, dry sliding and fretting) are used as reference and six key engineering materials are featured: the AISI 52100, the SAE 305, a 60:40 PC/ABS blend, a DLC, the 7050-T451 aluminium alloy and the Ti-6Al-4V alloy. Results show the advantages and liabilities of each solution scheme while emphasizing the inherent challenges of wear modelling.

1. Introduction

The contact between hard surfaces is an intrinsic part of manufacture processes. The repetitive movement of tools, over billets or metal sheets, usually enables several surface degeneration mechanisms. For instance, the failure mode observed in cutting, rolling and sliding systems is often related to a combination of localised inelastic strains, followed by damage and abrasive wear. Small amplitude relative motion is also commonly responsible for the development of fretting wear at the interface of structural components and moving parts. Moreover, the material loss removes protective coatings and surface treatments which catalyse undesired features such as roughness increase, corrosion and stress concentration. In order to evaluate the effects of wear, a significant number of formulations have been introduced.

1.1. *Wear formulations*

For several decades now, it has been known that wear is directly related to the relative motion observed at the interface, the material properties of the contacting bodies and the characteristics of their surfaces [1]. Moreover, many mathematical expressions have been presented to address this phenomenon. The most prevalent amongst them, Archard's law [2], defines a (dimensionless) wear coefficient, K, which is directly related to the worn volume, V, the hardness of the softest surface, H, a so-called "light" normal load, W, and a sliding distance, L, as follows:

$$K = VH/WL. \tag{1}$$

Therein, a Hertzian condition [3, 4] is assumed, meaning that the deformations are regarded as small and entirely under the elastic regime. Moreover, one of the solids must be considerably harder (most commonly a "tool") than the counterpart being worn and the hardness of this counterpart must not change as it is worn. Therefore, this method is particularly suited for steady-state wear processes and widely used in experimental analysis. For instance, in the design of dry rubbing applications it is common to define the component's lifetime, t, such that

$$t = \frac{w}{P\vartheta}\frac{H}{K}, \tag{2}$$

where w is the wear depth, P is the nominal pressure and ϑ is the sliding velocity. The material constants, H and K, are commonly considered as constants. However, they can also be rewritten as time-dependent variables, $H(t)$ and $K(t)$, which can be updated during the wear process. Moreover, this approach can be redefined in terms of a specific wear, k, that can be obtained at any given normal load, N,

$$k = K/H = V/NL. \tag{3}$$

In this interpretation of the problem, the effects of wear resistance and surface hardness are contained in a dimensional wear coefficient which again can be regarded either as a constant, k, or a time-dependent variable, $k(t)$.

Another approach is obtained by defining the product between the load and the sliding distance as work or, more precisely, as dissipated energy [5]. Hence, Eq. (3) can be rewritten as

$$\alpha = V/\sum_{i=1}^{n}\Pi_i, \tag{4}$$

where the average energy-wear coefficient, α, is defined as the ratio between the worn volume, V, and the dissipated energy, Π, accumulated throughout a total of n sliding cycles, i. This approach, first presented in [6], leads to an even more self-contained wear rate parameter which can be easily adapted to wear problems with varying loading conditions. For instance, in a problem with multiple sources of nonlinearity leading to an unstable behaviour, the wear process can be investigated at individual sliding cycles, such that

$$\alpha_i = V_i/\Pi_i, \quad i = 1, \ldots, n. \tag{5}$$

Meaning that, at every stage of the wear process (e.g., running-in, elastic strains, inelastic strains and coating-substrate transition) the energy-wear coefficient can be precisely calibrated using controlled experiments.

A comprehensive set of wear equations, each one specially designed for specific settings, has been derived from these primary concepts [7]. However, the initiation and evolution of wear in complex industrial problems is generally difficult to predict using mathematical equations. They often encompass many sources of nonlinear behaviour such as changes in size, shape and roughness of the contact area, transition between strain regimes and accumulation of debris or formation tribofilms in the interface. For this class of problems, numerical modelling is often a more efficient solution method.

1.2. *Wear modelling*

To model wear one must first enforce contact constraints in both normal and tangential directions. Moreover, a definition of wear resistance and a method for the topography update must be introduced. Several frameworks addressing these three building blocks have been proposed. For instance, solutions for quasi-static problems at small strains can be found in [8, 9] while structural dynamic contact problems are addressed in [10]. Frameworks for reciprocating-fretting sliding have been proposed in [11, 12] and a continuum framework for the analysis of finite wear quasi-steady-state problems has been introduced by [13].

Herein, two different frameworks for the solution of wear problems are featured. The first framework, presented in [14], consists of an in-house code built using python™ scripts for input/output data and FORTRAN subroutines for the remaining procedures. It employs an implicit scheme and the dissipated energy method. The contact constraints are fulfilled via the mortar segmentation method coupled with dual shape functions for the Lagrange multipliers [15]. This scheme yields an accurate description of force-displacement at the contact zone. However, it requires the consistent linearisation of the constraints and a Primal-dual active set strategy for the solution of the global stiffness equation system [16]. Moreover, this framework is intricate and difficult to implement when compared to traditional methods. The second framework, described in [17], employs the ABAQUS Explicit FEA (v6.14) by Dassault Systèmes® as a backbone, an UMESHMOTION routine (written in FORTRAN) for wear evaluation using Archard's method and python™ scripts for input/output data. This framework features an explicit central-difference time integration scheme and the standard segment-to-segment Lagrangian contact enforcement method. Therefore, it entails robust and easy-to-implement methods. Nevertheless, it is expected to be a less accurate representation of contact forces and wear effects. In both approaches, the wear is considered as an immediate consequence of frictional contact and its effects are evaluated as soon as this condition is achieved.

2. Comparative Examples

The problems described here were carefully selected to display the strengths and liabilities of the solutions schemes mentioned in Section 1.2. Indicators such as memory requirement and computational time are used to evaluate the performance of the numerical frameworks. Experimental data, retrieved from previous works (references provided in each example), are used as reference to assess the evolution of wear volume in terms of the relative sliding distance and the normal contact force.

Profiles of the wear scars are obtained using a Confocal Laser microscope and a scanning electron microscope (SEM), representative images of the profile measurements can be found in Fig. 1. Focus is given to three sliding regimes: fretting, rotational abrasion and reciprocate dry sliding.

2.1. *General settings*

Due to the geometrical characteristics of the three problems, a two-dimensional setting can be assumed. The first problem is axisymmetric (z-axis), while the second and third problems can be defined in a plane-strain state. The wear volume computed by the numerical frameworks is obtained using geometrical functions defined in terms of the dimensions of the wear scar (wear depth, wear length, wear radius, etc). Both the local

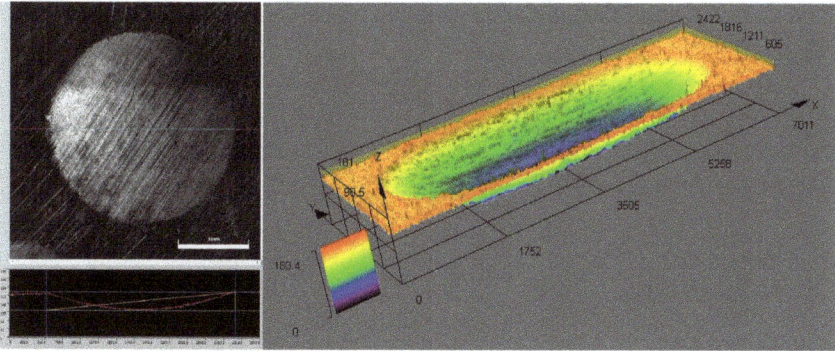

(a)   (b)

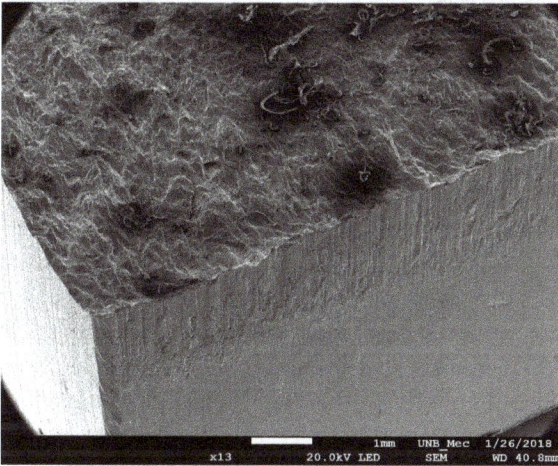

(c)

Fig. 1:   Representative microscopic profiles of contact zones: (a) confocal laser (surface view); (b) confocal laser (profile view); (c) SEM (top-view of fractured zone and side view of wear scar).

contact coordinate system (x/a) and the Cartesian coordinate system (x, y) are defined at the initial contact point between the two solids.

In order to enforce consistency across all numerical models, a few general considerations have been made. For instance, all solids are discretised using 4-noded linear quadrilateral finite elements with full Gaussian integration. The size of contact elements has been set to $10\,\mu$m while the maximum size of the internal elements is equal to 1 mm. In order to ensure the good performance of the topography update scheme, a structured mesh is

Table 1:   Mechanical properties of the materials employed in the experimental tests and simulations.

Properties	AISI 52100	SAE 305	PC/ABS	DLC	7050-T451	Ti-6Al-4V
Density, $\rho(\text{kg/m}^3)$	7810	2450	1140	3010	2830	4430
Poisson's ratio (ν)	0.29	0.33	0.35	0.22	0.33	0.29
Young's modulus, $E(\text{GPa})$	210.0	69.1	2.2	87.2	73.4	119.4
Yield stress, $\sigma_y(\text{MPa})$	890	305	54	900[a]	454	850
Strength coefficient, $K(\text{MPa})$	1450	206	—	—	400	1410
Strain hardening exponent, n_H	0.07	0.20	—	—	0.17	0.28

Note: [a]Value obtained considering a Yield by cleavage and the following relationship: $H/\sigma_y = 0.07 + 0.60 ln(E/\sigma_y)$ for a measured Hardness, H, of 2545MPa.

employed in the wear zone. A second global definition is that all solids are regarded as deformable bodies and their mechanical behaviour is presumed to be inelastic. A selection of six key engineering materials is featured: (1) the AISI 52100 steel, used in manufacturing tools; (2) the SAE 305 aluminium, employed in components of electric power transmission lines; (3) a 60:40 blend of polycarbonate/acrylonitrile-butadiene-styrene (PC/ABS), used in automotive parts, appliances and household items; (4) a diamond-like carbon (DLC) coating that promotes a significant increase in the wear resistance of soft substrates; (5) the 7050-T451 aluminium alloy, widely used in aircraft components; and (6) the Ti-6Al-4V alloy, an extremely resistant material commonly found in biomechanical implants and in aerospace components. Their relevant properties are listed in Table 1.

The constitutive model chosen for the representation of the metallic materials is the classical von Mises model [18], which is defined in terms of the stress deviator, **s**, and assumes the onset of plastic yield when the equivalent stress, σ_{eq}, reaches a critical value, σ_y,

$$\Phi^p = \sigma_{eq} - \sigma_y = \sqrt{\frac{3}{2}\mathbf{s} : \mathbf{s}}. \tag{6}$$

Moreover, the work hardening is defined in terms of the Ludwik–Hollomon equation,

$$\sigma = \sigma_y + K \left[\epsilon_p\right]^{n_H}, \tag{7}$$

where K is the strength coefficient, ϵ_p is the plastic strain and n_H is the strain hardening exponent. The PC/ABS and the DLC are regarded as a perfectly plastic and a brittle material, respectively.

All simulations have been performed on a Intel® Xeon® E5-2630 v3 CPU (8 cores, 3.2 GHz, 20 Mb cache) while using up to 128 GB of DDR4 memory at 3200 MHz. The convergence criterion is set as the residual tolerance $\mathcal{E}_r \leq E - 10$.

2.2. *Rotational abrasion*

The first case study features a sphere (10 mm radius) set to rotate over a flat disc (5 mm radius and 4 mm thickness). The sphere is made of AISI 52100 while two materials have been chosen for the flat disc counterpart: the SAE 305 and the PC/ABS 60:40. A schematic representation of the problem is given in Fig. 2. The faded regions and loads are disregarded in this analysis as they do not affect the contact interface.

The sphere and the flat disc are placed in contact at an 80° angle in relation to the horizontal plane. This position favours a constant rotation axis (no wobbling or vibration) while also allowing for a continuous feed of abrasive medium to the interface. The presence of the abrasive medium ensures a wear mark shaped as a smooth spherical cap, which is an ideal condition for comparison to a finite element model. The placement angle also promotes the vector decomposition of the sphere's weight such that a contact force, P, equal to 0.11N is maintained. It's important to empha-size that this "light" normal load enables only elastic strains. Therefore, a

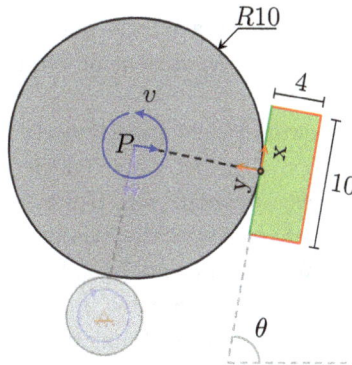

Fig. 2:   Rotational abrasion: schematic representation of the sphere (Gray, AISI 52100) on disc (Green, SAE 305 or PC/ABS) setup. Red lines highlight the restrained sections. Dimensions in mm.

low computational cost is expected. A velocity, v, of 0.1 m/s is applied to the sphere's outer surface and five predefined sliding distances have been selected for analysis: 25 m, 50 m, 75 m, 100 m and 125 m.

The friction coefficient assumed for the steel/aluminium lubricated interface is equal to 0.35 and, as reported in [19], the measured wear coefficient for this interface is 4.156E-10 mm^3/Nm while for the steel/polymer these values are 0.22 and 8.262E-5 mm^3/Nm, respectively.

The numerical model adopted in this example is depicted in Fig. 3. The sphere contains a 1-mm thick annular partition which is used to control the number of finite elements in the contact zone. A 4-mm-multi-1-mm partition is defined in the flat counterpart. This structured region is regarded as the wear box (contact candidates able to model the wear effect). The solid's discretisations are performed using 380806 elements for the sphere and 51661 for the flat disc.

The simulation is divided into a contact phase and a rotation phase. The contact phase comprises the weight load application in 10 equal increments. Each sphere's rotation is equivalent to a sliding distance of 62.8 mm which is modelled in 180 angular increments of 2°. In order to achieve the five targeted total sliding distances, the following number of rotations have been performed during phase two: 398, 796, 1194, 1592, 1990.

The experimental results and the numerical predictions for the total wear volume are given in Fig. 4. It is important to emphasise that the SAE 305 has a wear resistance almost 200 times higher than the one of the PC/ABS. Therefore, different units of measurement have been selected in order to plot both materials in a single graph. Since the contact force is kept constant throughout the tests and the sliding distances are predefined, the experimental wear volume data (circle markers) follows an expected linear trend (dashed line) with a coefficient of determination, R^2, equal to approximately 0.99.

Both numerical frameworks yield precise wear volume predictions for the SAE 305, at almost all sliding distances considered. The exceptions are the second and fourth measurements, which are found out of the trendline. Taking the trendline as reference, the average relative errors observed are 6.8% for the Explicit framework and 7.7% for the Implicit framework. However, the results for the PC/ABS show a deterioration of the explicit framework wear prediction as the total sliding distance increases. The average relative errors observed for the PC/ABS results are 9.9% for the Explicit framework and 2.8% for the Implicit framework. This loss in accuracy of the explicit wear prediction can be attributed to the accumulation of

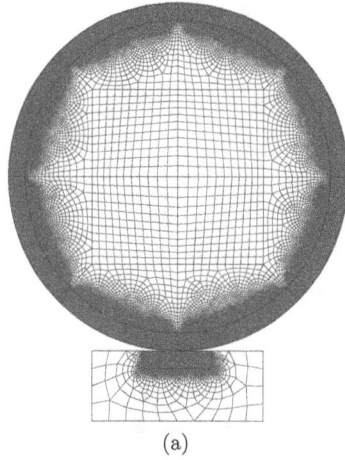

(a)

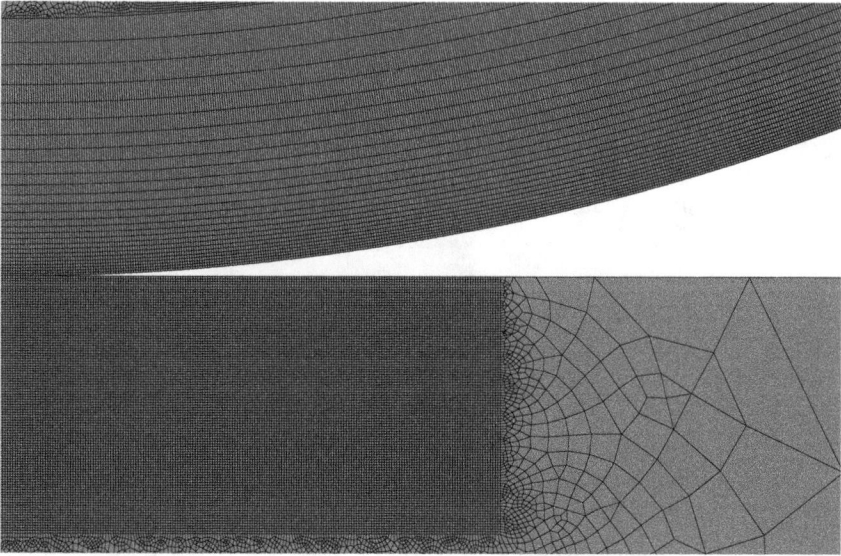

(b)

Fig. 3: Rotational abrasion: finite element discretisation. (a) Full view and partitions; (b) close view of the contact zone.

error while evaluating the normal contact force distribution and the contact area throughout the simulation. For instance, the maximum contact force, $F(x/a = 0)$, which is a reaction to the sphere's weight and takes place at the contact's zone center, has shown a reduction of 5.9%. The value dropped from the initial 0.110N to 0.103N at the end of this simulation. This can

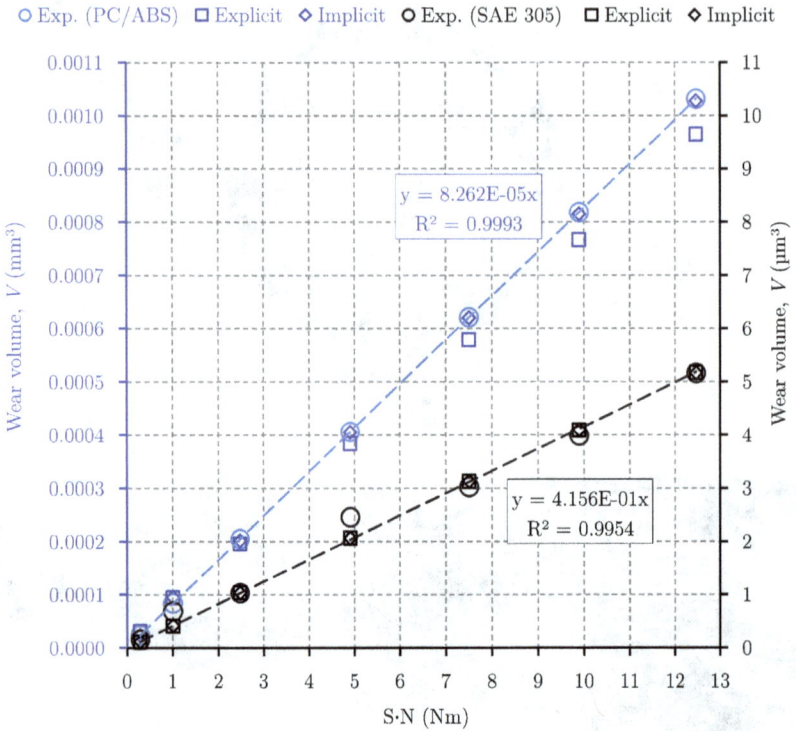

Fig. 4: Rotational abrasion: Evolution of the wear volume, V, vs. the sliding distance, S, multiplied by the normal load, N.

only be explained by a computational error during the evaluation of the contact area and the enforcement of the contact constraints.

The initial contact length, $2a$, for the steel/aluminium and the steel/polymer interfaces are $48\,\mu m$ and $139\,\mu m$, respectively. Both frameworks produced similar values. As the contact surfaces have a fairly small size (around 100–200 initial contacting elements) when compared to the dimensions of the solids involved, the computational cost remained relatively low. Moreover, despite the fact that the material loss of the PC/ABS is much more pronounced than the one of the SAE 305, both materials have shown a similar computational cost. Therefore, only the data related to the SAE 305 is detailed in Table 2.

The explicit framework requires 20% less memory and it takes 10% less computing time per iteration. However, it requires 59% more iterations per time-step to achieve the equilibrium/convergence. This leads to a processing 31% slower than the implicit framework. To put it in perspective, the total

Table 2:   Rotational abrasion: Framework performance comparison.

	Explicit	Implicit
Memory required for eq. system assembling (GB)	5.2	6.5
Average number of iterations per time-step	6.5	4.1
Average number of iterations per rotation	1070	738
Average computing time per iteration (ms)	18	20
Average computing time per rotation (s)	19.26	14.76

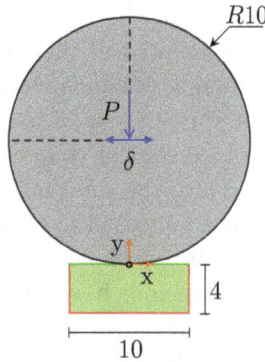

Fig. 5:   Reciprocate sliding: schematic representation of a sphere (Gray, AISI 52100) over flat disc (Green, SAE 305 or DLC) setup. Red lines highlight the restrained sections. Dimensions in [mm].

computing time of the whole simulation (1990 rotations required for a 125 m sliding distance) is: 10 h 38 min 47 s for the explicit framework and 08 h 09 min 32 s for the implicit framework (difference of approximately 2.5 h). Nevertheless, these results indicate that — as long as the loading conditions are properly modelled using adequate increments of a load that is limited to the formation elastic strains only — both methods are able to provide an accurate representation of the abrasion wear phenomenon at a reasonable computational cost.

2.3. *Reciprocate sliding*

In the second study case, a variation of the first one, the sphere (AISI 52100, 10 mm radius) has now its rotation axes fully restrained while being dragged in a reciprocate sliding motion over the flat disc sample (5 mm radius and 4 mm thickness). Two materials are considered for the flat counterparts: the SAE 305 and the DLC. This setup is depicted in Fig. 5.

The displacement stroke, δ, and frequency are set to 4 mm and 8 Hz. Two test durations are chosen: 30 min and 60 min. This loading condition leads to total displacements of 57.6 m and 115.2 m, respectively. The normal load, P, is set to 10 N which leads to substantial increase in contact area. The initial contact length, $2a$, is equal to 218 μm for the steel/aluminium interface and 208 μm for the steel/DLC pairing. In this dry frictional setting, deeper wear marks on the counterpart's surface are observed. As reported in [20], the measured values of friction coefficient and wear coefficient are 0.45 and 1.212E-3 mm^3/Nm for the steel/aluminium and 0.25 and 6.638E-6 mm^3/Nm for the steel/DLC pairing.

The finite element model adopted in this analysis is shown in Fig. 6. The mesh adopted in this example is similar to the previous one, but two changes have been made for the analysis of a reciprocate sliding setting. First, the size of the wear box on the flat counterpart has been enlarged to a 6 mm-multi-1 mm section. Thus, the total number of elements employed in the flat counterpart has been increased to 66,460. Second, the sphere no longer requires a refined mesh in its whole perimeter. Therefore, only the two partitions close to the contact zone have been refined. This solid is now discretised using 96,238 elements.

The simulation is divided into a normal load phase (divided into 10 equal increments of 1N) and a reciprocate motion phase where each displacement stroke is divided into 40 increments of 0.1 mm. The total number of cycles required to model the entirety of the reciprocate motion phase is equal to 14,400 for the 30 min test and 28,800 for the 60 min test. The results for the wear volume are depicted in Fig. 7.

The numerical predictions, from both frameworks, show quite similar values for the DLC wear volume and are also in good agreement with the experimental data. The average relative error for the explicit and implicit results is equal to 7.7% and 4.5%, respectively. Nevertheless, signs of an increasing deviation are once again observed in the explicit results. In fact, from the lower SN value (30 min, 57.6 m) to the higher SN value (60 min, 115.2 m) the relative error almost triplicates: 4.3–11.1%. The error for the implicit framework remains reasonably the same, 3.6% and 5.3%. The results obtained for the SAE 305 confirm this trend. Again, the explicit framework predicts wear volumes lower than expected which become less precise as the SN values increase. Since the contact area is more significant now, an initial error in its evaluation will have a larger impact on the evolution of the wear scar. At 57.6 m the explicit framework shows a relative error of 7.3% while at 115.2 m the error reaches 19.6%.

(a)

(b)

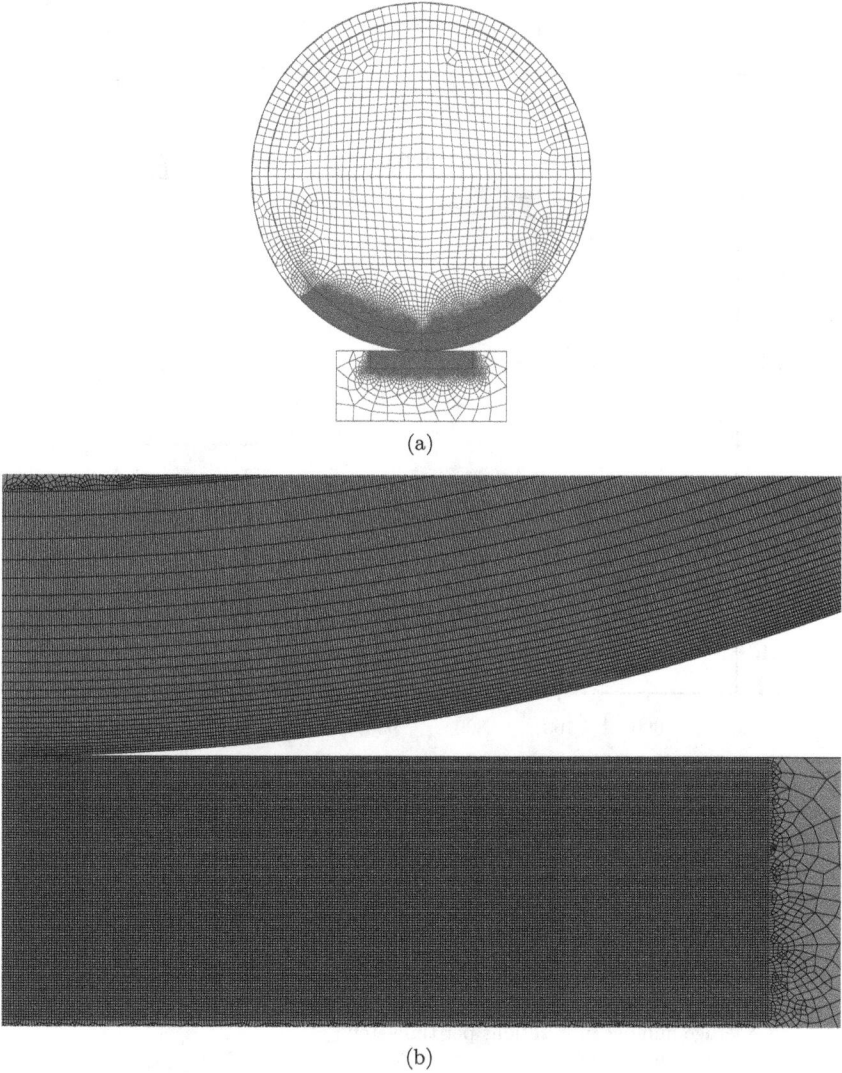

Fig. 6: Reciprocate sliding: finite element discretisation. (a) Full view and partitions; (b) close view of the contact zone.

The performance comparison of the two frameworks during the reciprocate sliding phase is given in Table 3.

When compared to the previous example, only a small increase in the memory requirement is observed. This additional memory is attributed to the increase in contact area which leads to a more costly segmentation

O Exp.(SAE 305)   □ Explicit   ◇ Implicit   O Exp. (DLC)   □ Explicit   ◇ Implicit

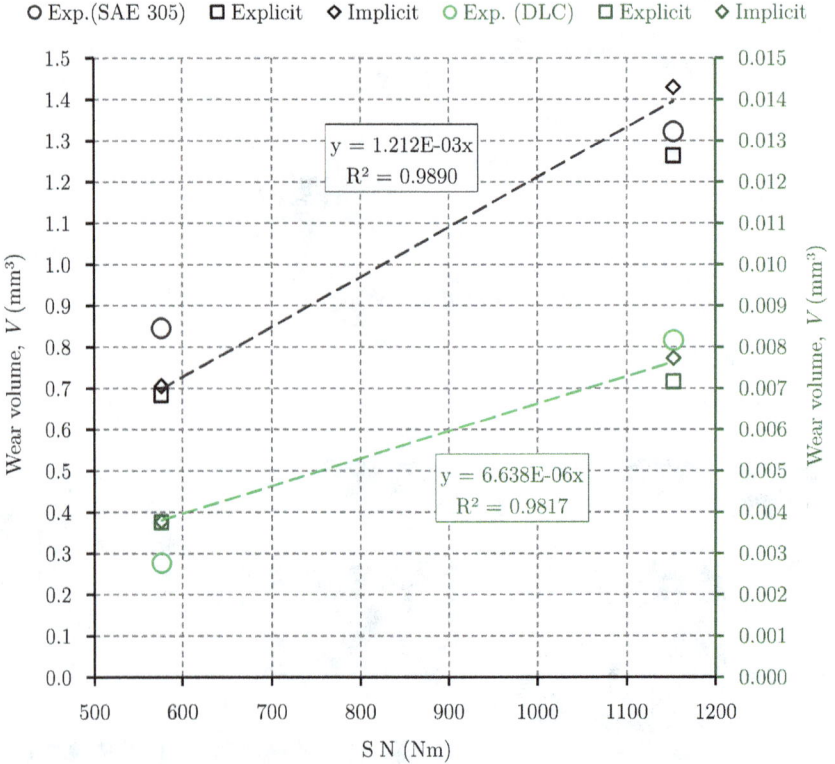

Fig. 7: Reciprocate sliding: Evolution of the wear volume, V, vs. the sliding distance, S, multiplied by the normal load, N.

Table 3:   Reciprocate sliding: Performance comparison.

	Explicit	Implicit
Memory required for eq. system assembling (GB)	5.3	6.6
Average number of iterations per time-step	6.5	4.2
Average number of iterations per cycle	520	336
Average computing time per iteration (ms)	21	24
Average computing time per cycle (s)	10.92	8.06

and constraint enforcement. A performance analysis shows once again the explicit framework as less (19.6%) memory expensive while being more (35.4%) time consuming. The time required for this simulation is considerably higher than the one observed in previous example. However, this condition is solely credited to the modelling of the reciprocate motion which required over 14 times more loading cycles. The total simulation time (60 min test condition) is equal to 3 d 15 h 21 min 36 s for the explicit and 2 d 16 h 30 min 43 s for the implicit scheme. Therefore, an absolute time difference of almost a day.

In summary, the results of this analysis indicate that the explicit formulation is unable to provide an accurate description of finite wear at high values of total sliding. This vulnerability can be linked to the STS method, which leads to a less precise evaluation of the contact area and contact forces. Moreover, its computational time far exceeds the one required by the implicit framework.

2.4. Fretting

The last case study entails the classical cylinder-to-flat contact configuration employed in fretting fatigue experiments. The schematic representation of this problem is given in Fig. 8.

A typical feature of fretting problems is that the contact zone is divided into a slip zone and an adhesion zone. Moreover, the sliding displacement is highly affected by the problem's stiffness (meaning they are material dependent). Furthermore, in some loading conditions the wear debris are trapped inside the contact zone which leads to variations of both friction and wear coefficients. Even further, both the pad and the flat counterpart are being worn. These conditions introduce severe challenges for the numerical model and some considerations had to be made. For instance, despite that wear effects are considered on both sides of the interface, only the wear on the

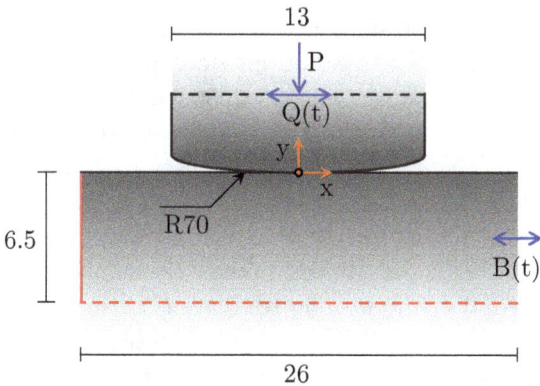

Fig. 8: Fretting: schematic representation of a cylinder pad on flat setup. Red line highlight the restrained sections. Dimensions in [mm].

Table 4: Fretting: Loading conditions employed in the Al 7050-T451 and Ti-6Al-4V experimental tests and numerical models.

	Al 7050-T451	Ti-6Al-4V
P (kN)	5.8	1.3
B_{max} (kN)	19.3	23.3
Q_a (kN)	2.2	2.2

flat surface is addressed here. This compromise has been made due to the difficulties inherent in measuring wear scars on the pad's curved surface. Materials and loading conditions that promote a stable tribological setting, with reasonably constant friction and wear coefficients, have been selected.

The materials chosen for the two interfaces of this study are: the Al 7050-T451 [21] and the Ti-6Al-4V [17, 22]. The experiments in question require the application of a constant normal load, P, a variable tangential load, $Q(t)$ and a variable bulk load, $B(t)$. The simulation is divided into two steps. Step one is the application of the normal load in 10 equal increments. Afterward, the normal load is kept constant. Step two is the simultaneous (no phase angle) application of $B(t)$ and $Q(t)$. The bulk load varies from 0 to B_{max} while $Q(t)$ varies from Q_{min} to Q_{max}, both are applied within 100 equal increments. This loading cycle is then repeated 50,000 100,000 and 200,000 times. The loading conditions for each material pairing are listed in Table 4.

A contact/wear zone (1.8 mm-multi-0.9 mm partition) has been defined in each side of the interface. A total of 19,257 elements are used in the discretisation of the pad while the flat specimen contains 21,432 elements. The flat specimen is restrained (x-direction) in its left side and symmetry condition is enforced at its bottom side. The friction coefficient and the wear coefficient assumed for the 7050-T451 interface are 0.60 and 2.951E-5 mm^3/Nm while for the Ti-6Al-4V interface these values are 0.50 and 1.430E-5 mm^3/Nm, respectively. The discretisation of the problem is depicted in Fig. 9.

The numerical results for the wear volume and the reference fretting test data are depicted in Fig. 10. Due to the large number of cycles, the wear volumes observed after the fretting tests are comparable to the two wear regimes previously analyzed. The testing conditions applied to both the aluminium and the titanium produced around 0.13–0.15 mm^3 of material loss. However, most of the wear volume is actually due to the size of the partial slip zone. The measured wear depth reached a maximum value of

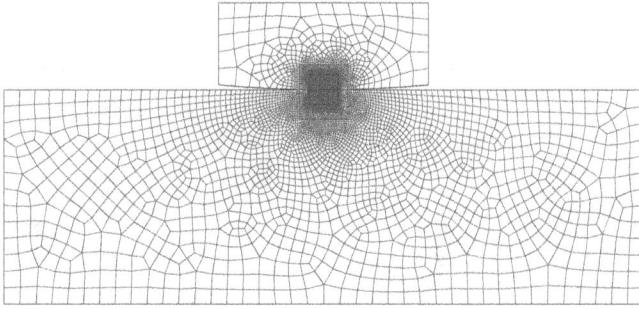

(a)

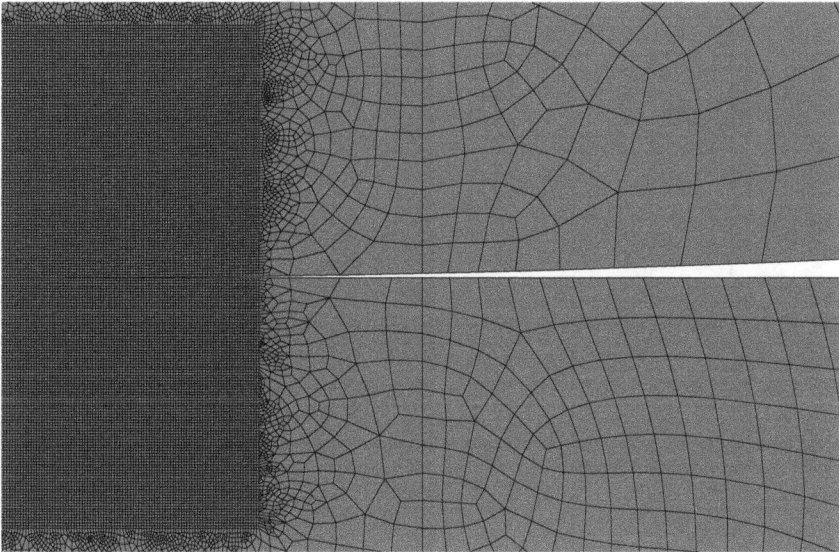

(b)

Fig. 9: Fretting: finite element discretisation. (a) Full view and partitions; (b) close view of the contact zone.

approximately 5 μm on the most damaged aluminium surface. A comparison between both frameworks shows again the implicit scheme as the more accurate description. When comparing the implicit framework predictions to trendline, the average relative errors are equal to 1.07% and 2.11% for the aluminium and titanium simulations, respectively. Moreover, the error remains fairly constant in each measuring point. Despite showing reasonable average errors (2.97% and 6.70%), the explicit framework displays once again a deterioration in precision at higher SN values. This issue is less

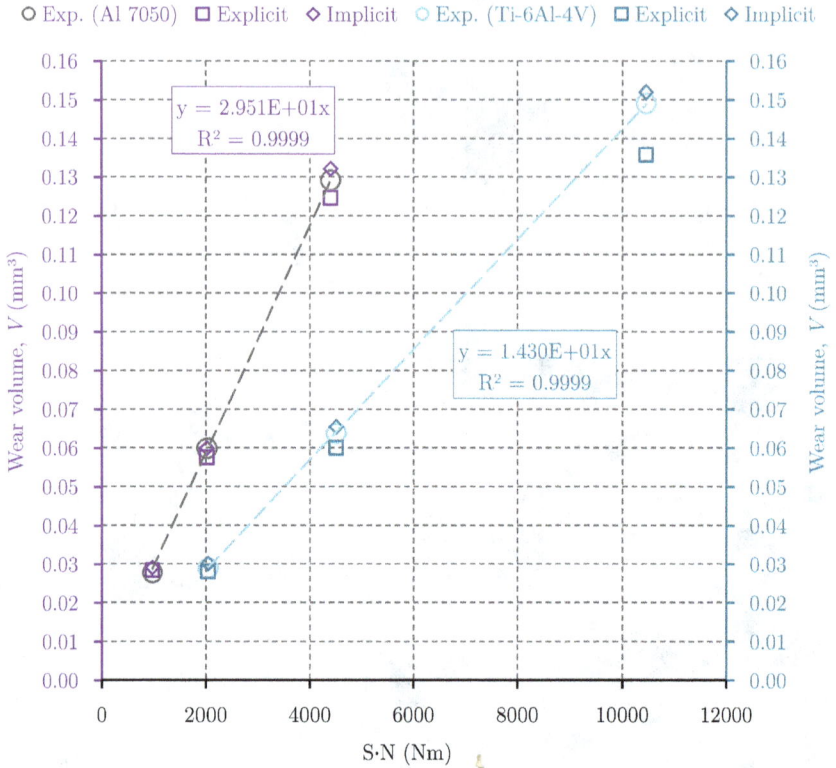

Fig. 10: Fretting: evolution of the wear volume, V, vs. the sliding distance, S, multiplied by the normal load, N.

pronounced in the aluminium tests (max error of 3.98%) but significant at the end of the titanium simulation (15.20%).

It is important to emphasise that in this particular problem both frameworks are stressed to their limits. Several attempts regarding mesh size, maximum number of iterations and increment cutting techniques have been tried before a successful run. The size of the load increments and the high number of increments per size are direct effects of the unstable behaviour this problem has shown. Regardless, the performance is considerably lower than the one observed in the previous examples as shown in Table 5. This behaviour is directly linked to the fact that variable loading and partial slip are present in this analysis. These two conditions constantly hinder the equilibrium state and the constraint enforcement while also requiring constant update of contact pairs during the segmentation procedure.

Table 5:   Fretting: Performance comparison.

	Explicit	Implicit
Memory required for eq. system assembling (GB)	3.7	4.2
Average number of iterations per time-step	7.1	8.3
Average number of iterations per cycle	710	830
Average computing time per iteration (ms)	42	69
Average computing time per cycle (s)	29.82	57.27

A performance analysis of the fretting simulations shows a different picture now. Since only forces are prescribed in this setup, the problems require either a dynamical approach or tiny load increments in order to retain a quasi-static condition. Therefore, the explicit framework has a clear advantage showing a 12.6% lower memory requirement and a more robust behaviour with less iterations per time-step. However, the most significant advantage is the average computing time per cycle which is 48% faster than the implicit framework. This difference becomes quite expressive when defined in terms of total processing time. The total simulation time (200,000 cycles) is equal to 69 d 00 h 40 min 02 s for the explicit framework while the time required for the implicit framework is 132 d 13 h 40 min 01 s. That's right, over two months more! The main problem of the implicit scheme is its sensibility to the iterative changes on the contact configuration while attempting to solve the force-equilibrium. These changes often require an update of the active set during the contact constraint enforcement which is followed by either a higher number of iterations per time-step or an increment cut followed by a reset of the convergence check.

3. Final Remarks

When comparing the two frameworks, it has been observed that the required memory for the assembling of the equations system plays a minor role in the performance of each solution. Most of the required memory is related to the global stiffness and the contact zone size. In the first two problems the stiffness influence is major, while in the third one the number of contacting elements produced a more balanced contribution. Nevertheless, all three examples can be easily run in a standard 16GB computer unit.

The computing time showed a strong relationship with the loading conditions. Prescribed displacement in a highly constrained problem has provided an ideal setting for the implicit framework while prescribed forces are far better handled by the explicit framework.

At lower levels of SN, both the explicit and the implicit frameworks have shown accurate wear predictions for the majority of the materials analyzed. The exception being the wear prediction for the titanium obtained with the explicit framework, which has shown higher relative errors from the beginning. At higher levels of SN, the explicit predictions are consistently worse. Again one exception has been observed: the wear predictions for the SAE 305 rotational abrasion problems which show an accurate description across the board. Evidence indicates that errors in the computation of the contact forces (specially the tangential force) and the length of contact are the culprits. However, additional research is needed to pin-point the exact reason behind those errors and how to correct them. Improving the constraint enforcement method may lead to a significant improvement in this method.

The great advantage of the implicit framework is its reliability. For all material setups and wear regimes, this framework has provided an excellent description of the wear volume evolution. However, its performance can be drastically affected by boundary constraints and loading conditions. The extension of this framework to better address time-dependent loads and dynamic behaviour would form an outstanding wear modelling approach.

References

[1] E. Rabinowicz, *Friction and Wear of Materials*. Wiley, New York (1965).
[2] J. F. Archard, Contact and rubbing of flat surfaces, *J Phys D Appl Phys.* **24(8)**, 981–988 (1953).
[3] H. Hertz, Über die berührung fester elastischer körper (on contact between elastic bodies), *Journal für die Reine und Angewandte Mathematik.* **29**, 156–171 (1882).
[4] K. L. Jonhson, *Contact Mechanics*. Cambridge University Press, Cambridge (1987).
[5] S. Fouvry, P. Kapsa, H. Zahouani, and L. Vincent, Wear analysis in fretting of hard coatings through a dissipated energy concept, *Wear.* **203–204**, 393–403 (1997).
[6] S. Fouvry, T. Liskiewicz, P. Kapsa, S. Hannel, and E. Sauger, An energy description of wear mechanisms and its applications to oscillating sliding contacts, *Wear.* **255**, 287–298 (2003).
[7] H. C. Meng and K. C. Ludema, Wear models and predictive equations: their form and content, *Wear.* **181–183**, 443–457 (1995).
[8] N. Strömberg, L. Johansson, and A. Klarbring, Derivation and analysis of a generalized standard model for contact, friction and wear, *Int J Solids Struct.* **33**, 1817–1836 (1996).

[9] C. Agelet de Saracibar and M. Chiumenti, On the numerical modeling of frictional wear phenomena, *Comput Methods Appl Mech Eng.* **177**, 401–426 (1999).

[10] N. Strömberg, A method for structural dynamic contact problems with friction and wear, *Int J Numer Methods Eng.* **58**, 2371–2385 (2003).

[11] I. R. McColl, J. Ding, and S. B. Leen, Finite element simulation and experimental validation of fretting wear, *Wear.* **256**, 1114–1127 (2004).

[12] S. Fouvry, T. Liskiewicz, and C. Paulin, A global-local wear approach to quantify the contact endurance under reciprocating-fretting sliding conditions, *Wear.* **263**, 518–531 (2007).

[13] J. Lengiewicz and S. Stupkiewicz, Continuum framework for finite element modelling of finite wear, *Comput Methods Appl Mech Eng.* **205–208**, 178–188 (2012).

[14] T. Doca and F. M. Andrade Pires, Finite element modeling of wear using the dissipated energy method coupled with a dual mortar contact formulation, *Comput Struct.* **191**, 62–79 (2017).

[15] B. I. Wohlmuth, A mortar finite element method using dual spaces for the lagrange multiplier, *SIAM J Numer Anal.* **38**, 989–1012 (2000).

[16] M. Gitterle, A. Popp, M. W. Gee, and W. A. Wall, Finite deformation frictional mortar contact using a semi-smooth Newton method with consistent linearization, *Int J Numer Methods Eng.* **84**, 543–571 (2010).

[17] R. A. Cardoso, T. Doca, D. Néron, S. Pommier, and J. A. Araújo, Wear numerical assessment for partial slip fretting fatigue conditions, *Trib Int.* **136**, 508–523 (2019).

[18] R. von Mises, Mechnik der plastischen formanderungvon kristallen, *Z Agnew Math Mech.* **8**, 161 (1928).

[19] V. Steier, M. S. T. Pires, and T. Doca, The influence of diamond-like carbon and anodised aluminium oxide coatings on the surface properties of the SAE 305 aluminium alloy, *J Braz Soc Mech Sci & Eng.* **40:41** (2018).

[20] M. S. T. Pires, T. Doca, V. Steier, W. M. da Silva, and M. M. Oliveira Júnior, Wear resistance of coated sae 305 aluminum alloy under dry friction reciprocate sliding, *Trib Letters.* **66:57** (2018).

[21] T. Gailliegue, T. Doca, J. Araújo, and J. Ferreira, Fretting life of the Al7050-T7451 under out-of-phase loads: numerical and experimental analysis, *Theor Appl Fract Mech.* **106** (2020).

[22] B. Ferry, J. A. Araújo, S. Pommier, and K. Demmou, Life of a Ti6Al4V alloy under fretting fatigue: study of new nonlocal parameters, *Trib Int.* **108**, 23–31 (2017).

Chapter 4

Wear in Heavily-Loaded Lubricated Contacts

G. E. Morales-Espejel

*SKF Research and Technology Development,
Meidoornkade 14, 3992 AE Houten, The Netherlands;
Université de Lyon, INSA-Lyon CNRS LaMCoS,
UMR5259, F69621, Villeurbanne, France*
guillermo.morales@skf.com

This chapter presents modelling strategies based on numerical and semi-numerical schemes to calculate local and average wear volumes in lubricated contacts of the elastohydrodynamic type. These contacts are present in heavily loaded machine elements like rolling bearings, gears and cam-follower couples. Local effects of wear are calculated when the roughness of the surfaces is of equivalent size as the lubricating film (mixed lubrication), thus wear modifies the microgeometry of the surfaces. Global effects are calculated when the profiles of the contacting elements are modified, so the overall pressure distribution and film thickness in the contact is modified.

1. Introduction

Heavily-loaded lubricated contacts are non-conformal contacts subjected to elastohydrodynamic lubrication (EHL) conditions. Like in the case of rolling bearings, gears and cam-follower mechanisms, present in any modern machinery. This kind of lubrication has different features compared with hydrodynamic lubrication present in conformal contacts, like sliding and journal bearings. Here the pressures are very high so elastic deformation of the surfaces is much larger than the film thickness itself; the high pressures also produce a dramatic increase of the lubricant viscosity due to the piezo-viscous effect and the compressibility of the lubricant also influences the pressures and film thickness [1, 2]. Wear in these conditions is difficult to predict since it depends on local lubrication conditions and potential asperity contacts, lubricant additives, temperatures and abrasive particles. However, with mild sliding conditions and full or mixed lubrication, wear is of the mild type, i.e. only local areas of the asperities will be modified. In the presence of abrasive particles, abrasive wear can modify also the surface profile, modifying the stress concentration points and under high loads and severe sliding adhesive wear can be important as a result of lubricant film collapse due to thermal effects [3].

1.1. *Wear mechanisms in EHL contacts*

There are several physical mechanisms that would give rise to mild-wear in EHL contacts. Wu and Cheng [4] give a good overview of potential physical effects in the removal of material in lubricated contacts. When asperities interact, material will be removed. The wear mechanisms for which this material is removed varies a lot depending on the asperity geometry, surface tribo-film, shear stress, lubricant chemistry and material properties. Even in conditions where lubrication is effective in completely separating the asperities, material is removed by the constant stress cycling and eventual fatigue. Also, tiny metallic particles in the lubricant will contribute with abrasive wear.

At low asperity contact temperature, the predominant mild wear mechanism is desorption. This happens in a system which is in the state of sorption equilibrium between bulk lubricant phase and the adsorbing solid surface (steel). When the pressure in the lubricant is lowered, some of the sorbed substance (reaction layer, steel) is transferred to the lubricant media. However, at high asperity contact temperatures, oxidative wear mechanism is more likely.

Abrasive wear in EHL contacts is also an important wear mode. Suspended abrasive particles in the lubricant are common after some running time. The particles can be transported to the EHL contacts via the lubrication systems even after filtering, the most tiny particles will remain. In case of unfiltered systems, e.g. gearboxes the number of abrasive particles (sand, steel, etc.) can be very important, so there can be a risk of high concentration of particles in the inlet of the contact. Triggering potential starvation and even other failure modes in the contact, like smearing [5]. Abrasive wear is more likely to actually change the profile of the contacting surfaces producing spots of high stress concentrations and accelerated fatigue [3].

2. Mixed-Lubrication Modelling

A complete solution for the problem of mixed lubrication (e.g. also known as partial lubrication) is up to now inexistent. It is still unknown how at the top of the contacting asperities the lubricant behaves or whether or not the lubricating film breaks to give way to tribolayer–tribolayer, steel–tribolayer, or steel–steel dry contact or simply becomes a patchy lubricating film at the molecular scale. Work is being done with molecular dynamics simulations to understand better how an extremely thin lubricating film behaves [6, 7]. Until a proper solution is found, only "engineering" models are available, which rely on strong hypotheses to be able to approximate the problem.

In the present section, it is not the intention of the author to address the basic physical phenomena of the partial-lubrication regime (e.g. film breakdown, lubricant micro-flow, droplet formation, surface energy effects, etc.), which are fundamental in the understanding and proper modelling of film formation and friction in this regime and would certainly prevent the author from the use of the Reynolds equation, among other things. The objective of the present section is only to present a fast engineering model capable of capturing the average behavior of different deterministic surface micro-geometries in the contribution of sliding friction and load carrying aspects of the lubricant and the asperities, once the friction coefficients in the boundary and full-film regimes are known. These kinds of model are known as "engineering" models.

The development of partial-lubrication engineering models was initiated in the 60s and early 70s. Pioneering experimental work on partial lubrication is due to Tallian *et al.* [8, 9]. Johnson *et al.* [10] used the stochastic asperity contact model from Greenwood and Williamson [11] to determine the load balance between the lubricant and the dry contacts. Although Johnson *et al.* were only interested in film thickness effects and contact time, the idea was widely used later by many authors for the calculation of friction in partial lubrication regime [12, 13]. Deterministic modelling of partial lubrication has been carried out by several researchers, e.g. Jiang *et al.* [14], Hu and Zhu [15], Redlich *et al.* [16], Wang *et al.* [17], and more recently Deolalikar *et al.* [18]. Here the model from Morales-Espejel *et al.* [19] will be described.

The approach used in [19] is based on the assumption of one nominally flat elastic–perfectly plastic rough surface in 'contact' with a smooth flat rigid counterpart. This approach allows for the simplification of the geometry and assumes that the mean film thickness and pressure are known (as in the central location of a contact). Therefore, only clearance and pressure fluctuations are calculated. The inlet and outlet of the contact are not fully included in the analysis. For the "dry" contact patches, a contact model with a fast Fourier transform (FFT) approach, as described in Stanley and Kato [20], is used but extended to three-dimensional (3D) geometries. Furthermore, the perfectly plastic behaviour of the asperities at high pressures is introduced. For the lubricated areas an upgraded version of the so-called "rapid methods" approach initially described for Newtonian fluids in [21] and later developed for non-Newtonian fluids in [22–24] will be used. Finally, the sharing of the load between asperities and lubricant is calculated by using the classical iterative approach of flow balance as described initially by Johnson *et al.* [10]. The final model avoids the expensive full numerical calculations from previous models. The normalised friction results calculated for real measured surfaces with the present model are compared with experimental measurements from a ball-on-disc machine. The results are encouraging and show that this simple approach is able to capture the overall behaviour of the sliding friction in the transition regime. Since thermal effects are not included, the model is limited to low sliding values, e.g. raceway contacts in rolling bearings.

2.1. *Dry contact solver*

Following the description in [19], for a given pressure distribution on a half-space $p(x, y)$, the surface elastic displacements $u(x, y)$ can be calculated using FFT by applying the following equation:

$$u = IFFT[w \cdot FFT(p)], \qquad (1)$$

where w is a matrix containing numerical factors and it is known as the frequency response function. For the elastic homogeneous problem, this matrix can be calculated as follows:

$$w(1, 1) = 0,$$

$$w(i, j) = \frac{1}{\sqrt{(i-1)^2 + [(j-1)l]^2}}, \quad \forall\, i = 1, \ldots, \frac{n_x}{2} \text{ and } j = 1, \ldots, \frac{n_y}{2},$$

$$w(i, j) = \frac{1}{\sqrt{(i-1)^2 + [(n_y - j + 1)l]^2}}, \quad \forall\, i = 1, \ldots, \frac{n_x}{2} \text{ and }$$

$$j = \frac{n_y}{2}, \ldots, n_y,$$

$$w(i,j) = \frac{1}{\sqrt{(n_x - i + 1)^2 + [(j-1)l]^2}}, \quad \forall\, i = \frac{n_x}{2} + 1, \ldots, n_x \text{ and}$$

$$j = 1, \ldots, \frac{n_y}{2},$$

$$w(i,j) = \frac{1}{\sqrt{(n_x - i + 1)^2 + [(n_y - j + 1)l]^2}},$$

$$\forall\, i = \frac{n_x}{2} + 1, \ldots, n_x \text{ and } j = \frac{n_y}{2} + 1, \ldots, n_y, \tag{2}$$

with $l = L_x/L_y$ and L_x, L_y represent the roughness lengths in x and y directions.

The contact problem is solved by finding the pressures that minimise the equivalent variational statement [25],

$$\min(f) = \frac{1}{2} \int_S pu\, dS + \int_S pg\, dS, \tag{3}$$

and

$$\frac{1}{S} \int_S p\, dS = p_{\text{target}}, \quad p \geq 0, \tag{4}$$

where f is the total complementary energy, g is the gap between the rigid plane and the undeformed elastic surface. Finally, p_{target} is the imposed prescribed load as a mean or target pressure.

2.1.1. *Algorithm*

The numerical algorithm employed to find the pressures from Eq. (4) is also described in reference [19]. Beginning with a guess for the matrix p which meets the equality and inequality constraints (a uniform pressure p_{target} is convenient), then

1. Calculate a candidate pressure matrix $p' = p - \text{grad}[f(p)]$. In general, p' will violate the constraints.
   $\text{grad}\,[f(p)] = u(p) + g$ for f quadratic.
2. Shift p' uniformly up or down so that the sum of the positive pressures equals the target load.
3. Truncate all $p' < 0$, Thus, p' meets all constraints.
4. Set $p = p'$, and repeat until convergence.

As described in [19], the model can accommodate plasticity by recalculating the deformations using a perfectly-plastic material and the von Mises stress criterion.

2.2. *Lubricated contact solver*

The considered Reynolds equation is:

$$\frac{\partial}{\partial x}\left(\frac{\rho h^3}{12\eta_x}\frac{\partial p}{\partial x}\right) + \frac{\partial}{\partial y}\left(\frac{\rho h^3}{12\eta_y}\frac{\partial p}{\partial y}\right) = \bar{u}\frac{\partial(\rho h)}{\partial x} + \frac{\partial(\rho h)}{\partial t}. \tag{5}$$

In this equation, h is the film thickness, p is the hydrodynamic pressure, ρ represents the lubricant density and η_x, η_y represent the lubricant equivalent viscosities in x and y directions for a non-Newtonian fluid, all these variables are depending on the x, y locations and the time (t).

Considering a sinusoidal waviness of amplitude r_a (for a sinusoidal wave, amplitude is the distance from top to bottom divided by 2) moving along x with a speed u_2, thus the roughness is described by,

$$\delta r = r_a \exp(i\omega_x x)\exp(-i\omega_x u_2 t)\exp(i\omega_y y), \tag{6}$$

where the frequencies $\omega_x = 2\pi/\lambda_x$ and $\omega_y = 2\pi/\lambda_y$. With λ_x, λ_y being the wavelengths of the sinusoidal roughness in x and y.

Provided that the amplitude of the sinusoidal roughness is low, the hydrodynamic pressures generated by the roughness will also be sinusoidal,

$$\delta p = p_a \exp(i\omega_x x)\exp(-i\omega_x u_2 t)\exp(i\omega_y y). \tag{7}$$

These pressures will generate elastic deformation in the roughness, corresponding to the following elastic displacements,

$$\delta v = v_a \exp(i\omega_x x)\exp(-i\omega_x u_2 t)\exp(i\omega_y y), \tag{8}$$

where the relationship between displacement and pressure amplitudes is given by

$$v_a = \frac{4p_a}{E'\sqrt{\omega_x^2 + \omega_y^2}}, \tag{9}$$

with E' is the combined elastic module of the two surfaces used in EHL.

The surface deformation is superimposed on the original roughness to produce the clearance variation inside the contact

$$\delta h = h_a \exp(i w_x x) \exp(-i w_x u_2 t) \exp(i w_y y), \tag{8}$$

where $h_a = r_a + v_a$.

The pressure variation will also modify the density of the lubricant,

$$\delta \rho = \rho_a \exp(i w_x x) \exp(-i w_x u_2 t) \exp(i w_y y), \tag{9}$$

where ρ_a is related to the pressure variation by,

$$\rho_a = \left(\frac{\rho}{B}\right) p_a, \tag{10}$$

B is the bulk modulus of the lubricant at a given pressure, $d\rho/dp = \rho/B$. Now, assuming that the products of fluctuations and products of derivatives can be neglected, since a major consideration in the present approach is that the amplitude of the clearance and pressure are small respect to the smooth pressures and film thickness. Thus the Reynolds Eq. (5) will be linearised. The solution of the linearised equation can be described with the addition of two components in clearance and pressures: the particular integral travelling with the speed of the rough surface u_2 and the complementary function describing the propagated wave originated at the inlet, travelling at the average speed of the lubricant $\bar{u}$. Here the approach developed by Hooke *et al.* [23] is followed.

2.2.1. *Particular integral*

Expressions for p, h and ρ can be added to the smooth contact and substituted back into Eq. (5), then the equation is simplified to:

$$\frac{\rho h^3 w_x^2}{12 \eta_x} p_a + \frac{\rho h^3 w_y^2}{12 \eta_y} p_a = i(u_2 - \bar{u}) w_x \left(\rho h_a + \frac{\rho h}{B} p_a\right), \tag{11}$$

from Eq. (11) the solution for p_a can be obtained,

$$\frac{p_a}{r_a} = \frac{\kappa E'}{4} \frac{iQ}{1 - iQ - iCQ}, \tag{12}$$

where $\kappa = \sqrt{w_x^2 + w_y^2}$, $C = hE'\kappa/(4B)$ and

$$Q = \frac{48(u_2 - \bar{u}) w_x}{E'h^3 \kappa \left[(w_x^2/\eta_x) + (w_y^2/\eta_y)\right]}. \tag{13}$$

With an Eyring fluid, $\dot{\gamma} = (\tau_0/\eta)\sinh(\tau/\tau_0)$ the effective viscosities are,

$$\eta_x = \frac{\eta}{\cosh(\tau_m/\tau)}, \tag{14}$$

$$\eta_y = \frac{\eta(\tau_m/\tau_0)}{\sinh(\tau_m/\tau)}, \tag{15}$$

where τ_m is the mean shear stress.

Equation (8) defines the particular integral wave for the clearances.

$$\delta h_{pi} = h_a \exp(i\omega_x x)\exp(-i\omega_x u_2 t)\exp(i\omega_y y). \tag{16}$$

2.2.2. Complementary function

Hooke *et al.* [23] suggest that for non-Newtonian fluids with rolling-sliding conditions, the complementary waves (pressure and clearances) will decay in amplitude as they propagate inside the contact. They suggest an exponential decay respect to the inlet $(x' = x + b)$ location. Since this wave propagates with the average velocity of the lubricant the wave number in x will be $\omega'_x \approx \omega_x(u_2/\bar{u})$. The waves decay exponentially with distance at a rate β, the amplitude is then expressed as

$$\delta h_c = h_c\exp(i\psi x')\exp(-i\omega_x u_2 t)\exp(i\omega_y y), \tag{17}$$

$$\delta p_c = p_c\exp(i\psi x')\exp(-i\omega_x u_2 t)\exp(i\omega_y y), \tag{18}$$

with $\psi = \omega'_x + i\beta$. Hooke *et al.* [23] give the following expression for ψ.

$$\psi = \omega'_x + i\frac{E'h^3\sqrt{\psi^2 + \omega_y^2}[(\psi^2/\eta_y) + (\omega_y^2/\eta_y)]}{48\bar{u}\left[1 + \left(E'h\sqrt{\psi^2 + \omega_y^2}/(4B)\right)\right]}. \tag{19}$$

Hooke *et al.* [23] calculate the complementary function amplitudes h_c and p_c by solving the problem with a perturbation technique for different wavelengths and then interpolating to the desired wavelength. However, in [19] a simplification of this procedure was introduced. The proposed modification for the complementary wave amplitude is fully described in [19], it is the result of a curve-fit of many numerical simulations with non-Newtonian (Eyring) fluids and rolling-sliding conditions, the model is summarised next.

First, the film thickness is related to the clearance fluctuations by $h = h_0 + \delta h$, where δh given as a function of the two components is $\delta h = A_d = A_c + A_{pi}$. Notice the use of the variable A (e.g. A_d, A_c and A_{pi}) instead of h (e.g. δh, δh_c and δh_{pi} given by the equations above) introduced for the clearance fluctuations at this point in order to be consistent with the "amplitude-reduction" nomenclature, e.g. [26]. Thus, the amplitude of the complementary ripple clearance at a given location x, y, and for every wave number can be calculated as

$$A_c = A_d - A_{pi}, \tag{20}$$

where A_c is the ripple clearance amplitude of the complementary wave, A_d is the overall ripple clearance deformed amplitude and A_{pi} is the ripple amplitude of the particular integral. For the calculation of the complementary ripple, Eq. (20) it is very important to remove this latest amplitude from the original overall deformed ripple, because as will be discussed later, this amplitude is responsible for the bending down of the complementary function at short wavelengths, as discussed in [27].

Now, the overall ripple clearance deformation amplitude A_d needed in Eq. (20) to obtain A_c can be estimated by following the approached given in [19]:

$$A_d = \frac{A_i}{1 + 0.15\nabla_{nn} + 0.015\nabla_{nn}^2}, \tag{21}$$

where A_i is the harmonic component amplitude of the initial roughness and,

$$\nabla_{nn} = 0.8\nabla(1 + S/2)^{0.1+0.5K}, \tag{22}$$

with

$$\nabla = \frac{\lambda_x}{b}\left(\frac{M}{L}\right)^{0.5}, \tag{23}$$

λ_x is the harmonic component wavelength of the initial roughness, b is the Hertzian semi-width along the rolling direction and M, L are the Moes' dimensionless parameters for point contact.

Finally,

$$K = 1 - \tanh\left(0.25\frac{|Q|}{\nabla}\right), \tag{24}$$

and Q being the amplitude attenuation parameter of Eq. (13).

For any particular roughness or micro-geometry, a FFT is applied to obtain the full spectra. The above equations for the particular integral and the complementary function are applied to each harmonic of the spectra. The two component results are added to obtain the final solution for pressures and clearance fluctuations.

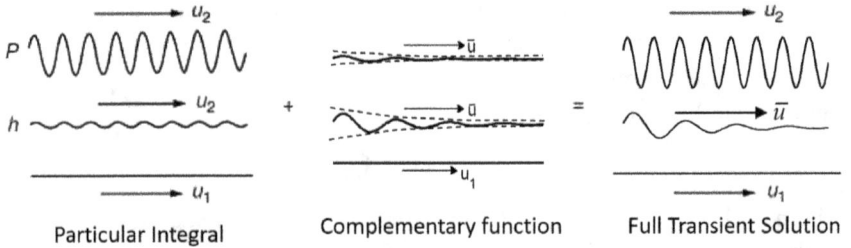

Fig. 1: Transient pressure and film thickness fluctuations that make the full transient solution for the rolling-sliding, non-Newtonian EHL, wavy surface contact, inspired [28], with the amplitude decay in the inlet disturbances as proposed by [29].

The combination of the two solutions is represented in Fig. 1 where the moving steady state solution represents the particular integral, the inlet disturbances are given by the complementary function equations.

2.2.3. *Applicability limits*

In [19] the limits of the applicability of the present approach are discussed. It is concluded that to ensure accuracy the pressure should everywhere exceed the value of Eq. (25).

$$
p\alpha \geq \ln\left[\frac{\pi p_1 h^3}{6\eta_0 \bar{u}\lambda_x \left[h - h^* \left(\frac{\rho^*}{\rho}\right)\right]}\right], \tag{25}
$$

where p_1 represents the combined (particular integral and complementary function) pressure amplitude assuming it is sinusoidal and h^* is the central film thickness with ρ^* the corresponding value for the density. However, it is also discussed in [19] that this condition is often not fulfilled and the solutions remains largely acceptable.

2.3. *The combined model*

The model for the transition from boundary lubrication to full-film conditions combines the dry contact model described above and the full-film model described in the previous section.

Here it is important to point out that it should be clearly understood that the Reynolds equation cannot be used to explain any film thickness break-down to produce dry contacts as long as the surface velocity remains different from zero and no lubricant-wall slippage is allowed. However, the present model only attempts to approximate the contact pressures and elastic deformation of the surface for the load sharing part that is important in friction. With this objective in mind, it is believed that in very small clearance locations, it might result in less error to calculate these parameters using a dry contact model rather than a continuum-mechanics lubrication model. However, this question remains to be resolved and it is a matter of ongoing discussions.

At very low speeds, the calculation of pressure, deformation and friction is more accurately done by using the dry contact model. On the contrary, at high speeds, when surely conditions of full film take place, this calculation is more accurately done with the full-film model. However, at intermediate speeds, the calculation of the load sharing between the full film and the dry contact models follows the following steps.

1. Surface and operation conditions are entered. The mean clearance h_{mean} is calculated with the use of any 'central film thickness' formula for smooth contact (if possible, a correction for roughness can be done). The mean pressure in the contact can be calculated in several ways. It can simply be the mean pressure in a Hertzian contact or if the roughness sample is small compared with the Hertzian contact width the maximum Hertzian pressure can be used p_h, performing the analysis at the centre of the contact. Here the latest is assumed.
2. The iteration process begins by assuming a load sharing fraction ϕ_{bl} for the dry contribution $(F_{\mathrm{dry}} = \phi_{\mathrm{bl}}F)$.
3. Using the load $F_{\mathrm{lub}} = (1 - \phi_{\mathrm{bl}})\,F$, solve the lubrication problem and find the pressures p_{lub} and clearances h_{lub}. A correction for the density is applied here. The mean clearance obtained from the film thickness formula incorporates the compression of the lubricant at high load. Depending on the proportion of the load carried by the lubricant film F_{lub}, the lubricant will expand. The mean clearance needs to be multiplied by the correction factor c_p obtained by inverting the Dowson–Higginson density law

$$c_p = \frac{0.59 + 1.34\phi_{\mathrm{bl}}p_{\mathrm{mean}}}{0.59 + \phi_{\mathrm{bl}}p_{\mathrm{mean}}}. \tag{26}$$

4. Evaluate the elastic-plastic, dry contact pressures p_{dry} and clearances h_{dry} using as input the lubricated clearances h_{lub} and the load F_{dry}.
5. The calculation of the transition clearances h_{tran} is done by assuming that in general long wavelength surface features have larger amplitudes that are more likely to make contact. Thus, the two clearances, lubricated

and dry, in the frequency domain (after FFT), say ($\tilde{h}_{\text{lub}}$ and $\tilde{h}_{\text{dry}}$) are compared for each frequency and the absolute maximum is selected. After the process, the IFFT is calculated to recover h_{tran}.

$$h_{\text{tran}_{(i,j)}} = \text{IFFT}\left[\max\left(\left|\tilde{h}_{\text{dry}_{(i,j)}}\right|, \left|\tilde{h}_{\text{lub}_{(i,j)}}\right|\right)\right]. \tag{27}$$

The evaluation in this manner avoids the use of unrealistically smoothen (otherwise use filtered) clearances as calculated by the full-film method in the far right end of the amplitude reduction curve. The clearances at very low speeds are to be limited by the dry contact clearances.

6. After having calculated the transition clearances h_{tran}, the corresponding pressures p_{tran} can be recovered by using the produced displacements and applying the inverse process to Eq. (1).

$$p_{\text{tran}} = \text{IFFT}[(w^{-1}) \cdot \text{FFT}(r - h_{\text{tran}})], \tag{28}$$

where $r - h_{\text{tran}} = u$. If plasticity has occurred, the pressures p_{tran} will have to be limited to p_{lim}.

7. The process continues until convergence for ϕ_{bl}, which is obtained once the corrected mean clearance equates the mean line of the deformed clearances h_{tran}. This is equivalent to assume that the load sharing equilibrium is reached when the volume of oil that goes through the contact ($c_p h_{\text{mean}}$) equates the free space in the deformed clearance h_{tran}, which is an assumption considered earlier in stochastic roughness models [10].

Figure 2 shows schematically how a sample of roughness of each surface can be followed by the mixed-lubrication model in a rolling-sliding condition where pressures and relative position of the roughness changes in time while going through the EHL contact. The actual calculation of pressures for two contacting rough surfaces in rolling/sliding movement under EHL and mixed-lubrication conditions is performed following the aforementioned partial lubrication model. However, here the procedure used to store and utilise those pressures in a time sequence is described in order to be able to calculate wear on the two surfaces. In the model, both surface topographies are followed by a window of analysis (Lagrangian flow specification) which moves throughout the contact following the Hertzian pressure profile, as shown in Fig. 2. Inside the window there is relative motion on the roughness in time (Eulerian flow specification), with changing local pressures as both surfaces move with the relative speed.

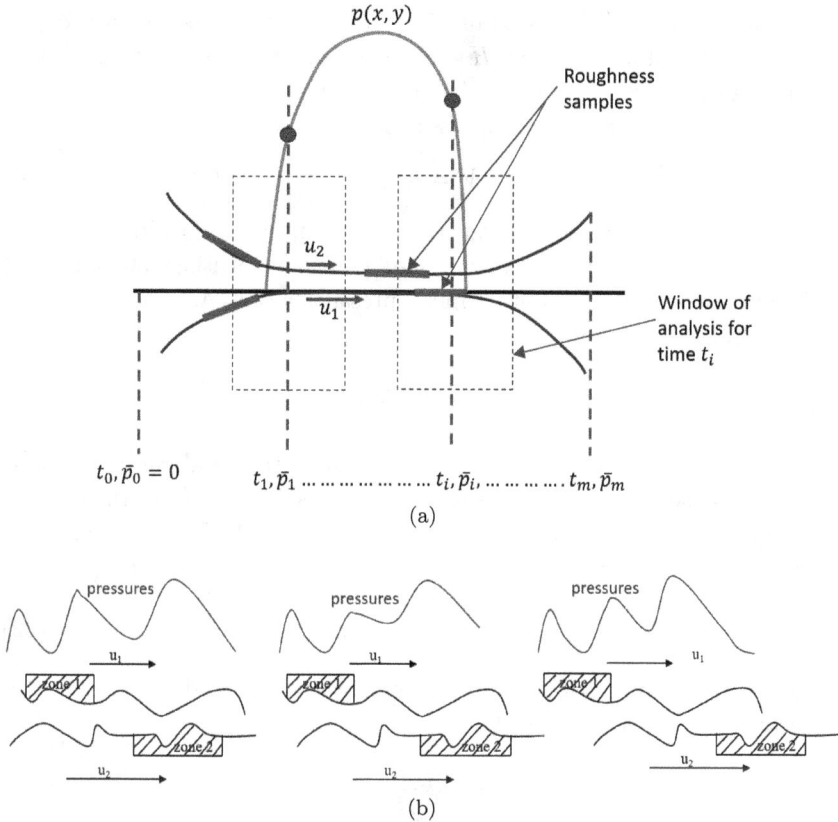

Fig. 2: Relative motion of two rough surfaces in a contact, as captured by the mixed-lubrication model (these figures were reprinted from [28] and [32] with permission from Taylor & Francis and Sage). (a) Roughness consideration in the model, with respect to the contact; (b) roughness details in rolling/sliding motion, case $u_1 < u_2$.

3. Mild Wear Model for EHL Contacts

Mild wear in EHL is understood as the wear that only modifies the asperities, it does not modify the profile of the contacting surfaces. It helps in the running in of surfaces. As mentioned in the introduction, there are several approaches to this [4], but here the model described in [30] will be pursued.

The original equation from Archard [31] for a single smooth contact is:

$$V = k\frac{F}{H}s, \tag{29}$$

$V =$ wear volume [m³] in a certain time, $k =$ dimensionless Archard coefficient [–], $F =$ contact force [N], $H =$ current surface material hardness [Pa], $s =$ sliding distance [m] in a certain time.

The wear volume can be expressed as:

$$V = hA_s, \tag{30}$$

where $h =$ removed surface layer thickness [m] in a certain time, and $A_s =$ sliding area [m²] in a certain time. Thus, substituting (30) into (29) and considering that the contact mean pressure $\bar{p} = F/A$:
$hA_s = k\frac{\bar{p}A}{H}s$, which gives:

$$h = k\frac{\bar{p}A}{HA_s}s. \tag{31}$$

Consider the wear per loading cycle (N), where the total contact time is t and it represents the time of passage of both sliding surfaces throughout the contact zone with the sliding speed, thus $A_s = A$. The removed layer thickness per number of cycles is:

$$\frac{h}{N} = k\frac{\bar{p}}{H}s/N. \tag{32}$$

Then, one can replace the sliding distance by $s = u_s t$ (sliding speed multiplied by the total contacting time):

$$\frac{h}{N} = k\frac{\bar{p}}{H}\frac{u_s t}{N}. \tag{33}$$

Consider now, that a new dimensional wear coefficient is defined as $k_p = k\frac{t}{N}$ [s], this wear coefficient is the k used in [32]. However, here k is redefined as the dimensionless wear coefficient, thus Eq. (33) can be rewritten as

$$\frac{h}{N} = k_p\frac{\bar{p}}{H}u_s. \tag{34}$$

Notice that further the total contacting time per loading cycle can be defined as

$$\frac{t}{N} = l/\bar{u}, \tag{35}$$

where l is the contact length ($\approx 2a$ for a Hertzian contact), so in fact Eq. (34) is also:

$$\Delta h = \frac{h}{N} = k\frac{\bar{p}}{H}u_s\left(\frac{l}{u_1}\right), \tag{36}$$

where u_1 is the speed of the analysed surface.

However, generalising Eq. (36) for a local definition of the parameters, but keeping constant speeds values, one has:

$$\Delta h(x,y) = k(x,y)\frac{p(x,y)}{H(x,y)}u_s\left(\frac{l}{u_1}\right). \tag{37}$$

And assuming that also the hardness is constant at asperity level (this may not be entirely true in reality, but for now it will be considered as a good approximation):

$$\Delta h(x,y) = k(x,y)\frac{p(x,y)}{H}u_s\left(\frac{l}{u_1}\right). \tag{38}$$

The local coefficient of wear $k(x,y)$ can accept either of the two values, k_{lub} or k_{dry}, depending on the local lubrication regime (for fully or boundary lubricated patches of the contact area, respectively) which can be obtained as follows. By evaluating the average value of the wear groove depth after a certain amount of cycles, on the relevant test rig, one can obtain the value of the total mean wear coefficient (with the use of the Archard's equation), which is always between k_{lub} and k_{dry}, depending on the lubrication regime. The mixed-lubrication solver calculates the shares of the lubricated and dry zones in the contact area which imply the contribution factors for k_{lub} and k_{dry}, respectively. Finally, assuming that $k_{dry} = 10k_{lub}$ (see [32]), one can obtain approximate values for the "dry" and "lubricated" coefficients of wear. The values of k_{lub} and k_{dry} have an obvious influence on the model results: the higher they are, the faster the evolution of the surface topographies becomes.

When it comes to the wear model used, there is no much more than Eq. (30). The difficulty now comes in the way this equation is applied in the contact of two rough surfaces and in the evaluation of the wear coefficient. The overall model flow chart is depicted in Fig. 3.

For the total local wear thickness $h_w(x,y,t)$ calculation in one load cycle, all the local $\Delta h(x,y,t)$ of the different time steps are averaged in time for each surface to obtain $h_w(x,y)$. In every load cycle, the relative motion of the roughness patterns is implemented in five time steps, which was found sufficient to obtain accurate results.

4. Example: Wear Calculation of Two EHL Surfaces with Relative Motion

An example is given on the evolution of surface topographies and profiles, with the operating conditions shown in Table 1, and presented in

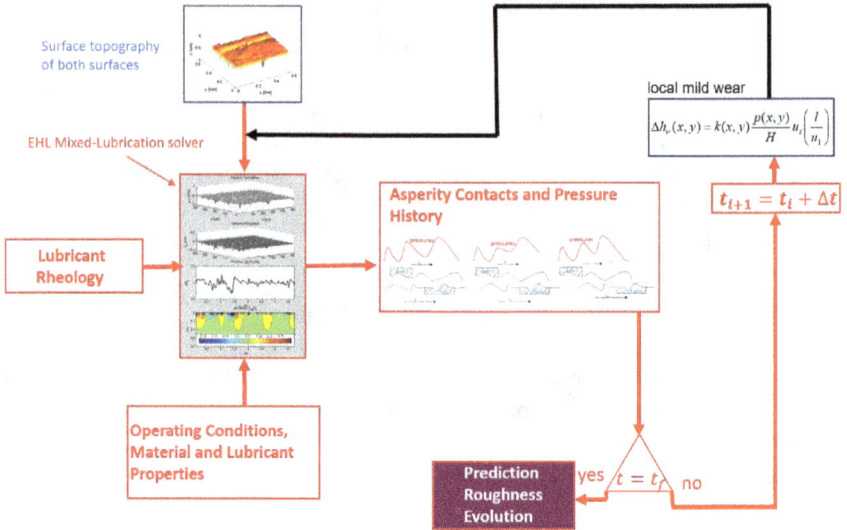

Fig. 3:   Complete EHL wear model flow chart, inspired from [30].

Table 1:   Contact data of model simulation.

η_0 mPa·s	α, GPa^{-1}	τ_0, MPa	E', GPa	S, –	$\bar{u}$ m/s	p_0, GPa	a, mm	b, mm	h_c, nm	R_{qi1}, nm	R_{qi2}, nm	Λ
8.1	27.8	3	206	0.02	1.0	1.5	0.125	0.52	86	48	191	0.44

Figs. 4(a) and 4(b). In this case, two surfaces with real (measured) rough-ness patterns — an initially rougher one ($R_{qi2} = 191$ nm, see Fig. 4(a), left) and a smoother one ($R_{qi1} = 48$ nm, see Fig. 4(b), left), come into a mutual rolling-sliding mixed-lubricated contact with initial value of the lubrication quality parameter $\Lambda = 0.44$. The lubricated and the dry coefficients of wear in this and in all the following examples of modelling results are: $k_{\text{lub}} = 3 \cdot 10^{-11}$; $k_{\text{dry}} = 3 \cdot 10^{-10}$, respectively.

The subsequent evolution of both topographies due to mild wear is presented in the middle (after $N = 2 \cdot 10^5$ cycles) and right-hand part (after $N = 10^6$ cycles) of Figs. 4(a) and 4(b). It can be concluded that the rougher surface (Fig. 4(a)) is gradually becoming smoother (one can see R_{q2} decreasing from 191 nm initially to 119 nm after $N = 10^6$ cycles), while the smoother one (Fig. 4(b)) first gets even smoother (R_{q1} decreases from

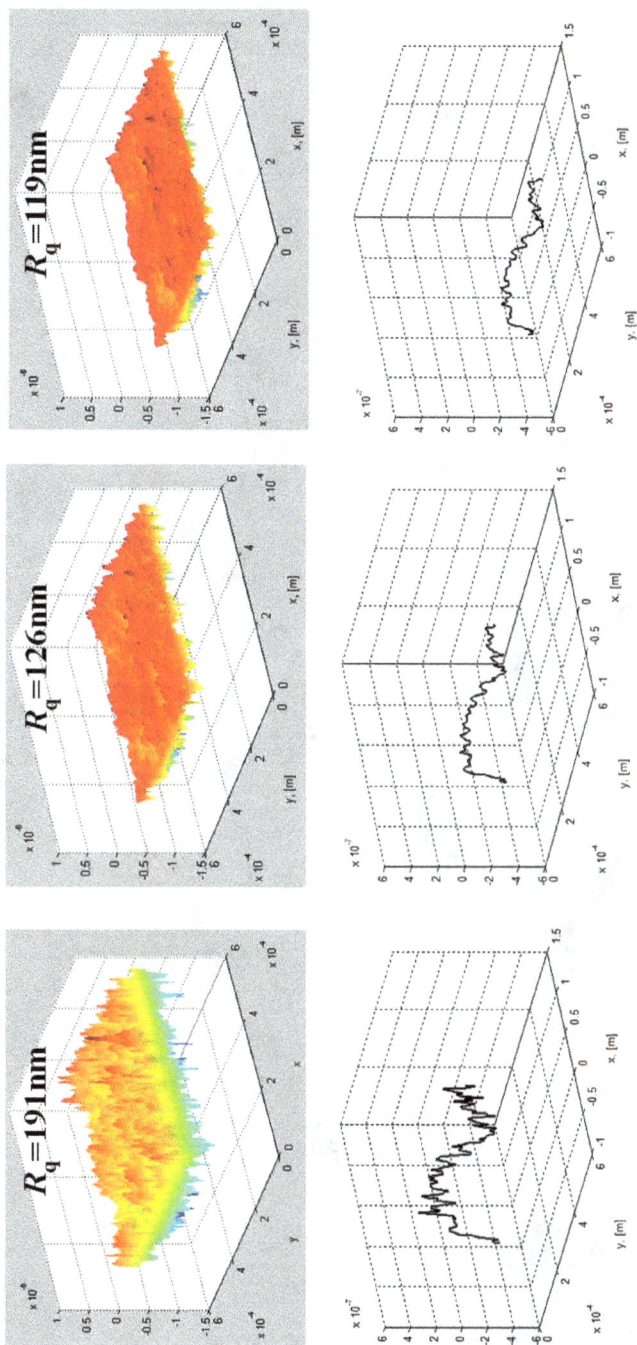

(a)

Fig. 4: Example of model results for roughness topography and profile evolution of both contacting surfaces: (a) rougher surface; (b) smoother surface. For all the results: left – initial; middle – after $N = 2 \cdot 10^5$ cycles; right – after $N = 10^6$ cycles; top – roughness topographies; bottom – evolution of the same roughness profile in the y-direction. The values of R_q are presented above the corresponding topographies. The operating conditions in this example are as follows: $R_{qi1} = 48$ nm, $R_{qi2} = 191$ nm, $p_o = 1.5$ GPa, $S = 0.02$, $U = 1$ m/s, $\Lambda = 0.44$, $k_{lub} = 3 \cdot 10^{-11}$, $k_{dry} = 3 \cdot 10^{-10}$. The two surfaces accumulate the loading cycles with identical rate. Figure reprinted from [30] with copyright is secured from SAGE.

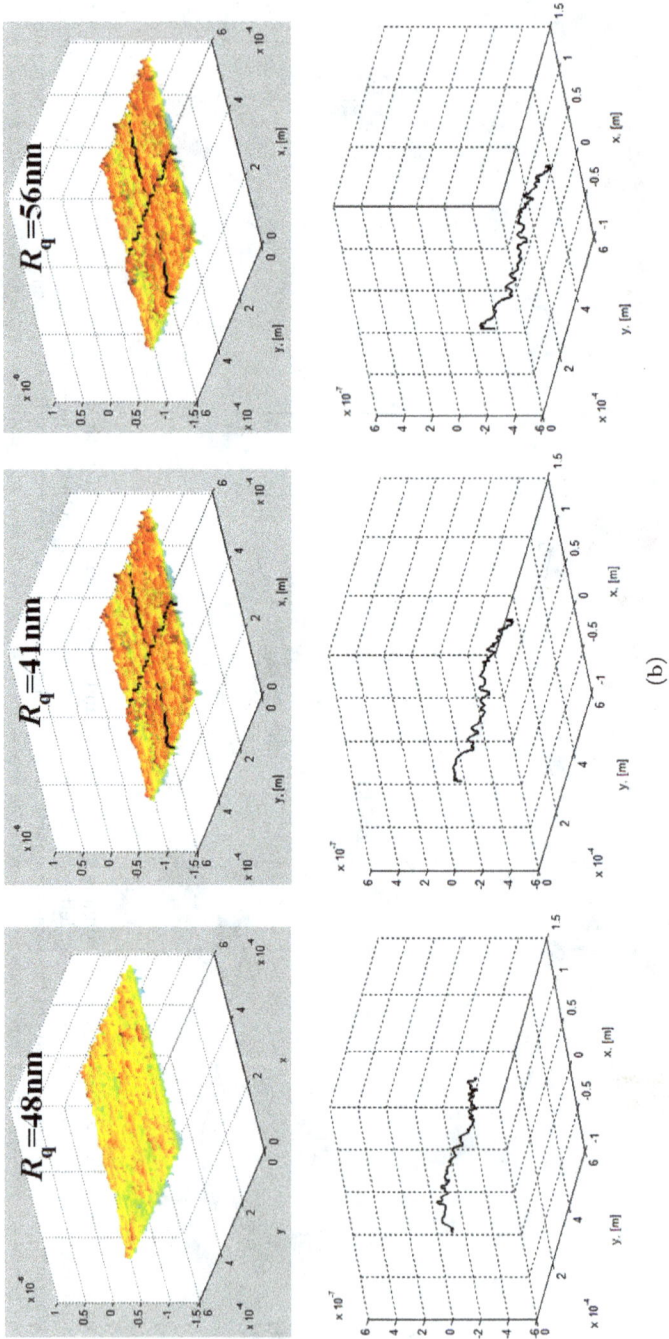

(b)

Fig. 4:   (*Continued*)

48 nm initially to 41 nm after $N = 2 \cdot 10^5$ cycles) but then it roughens (up to $R_{q1} = 56$ nm after $N = 10^6$ cycles). One of the possible reasons for this interesting behaviour is as follows. Initially, both surfaces start getting smoother due to mild wear which occurs (mostly) at micro-contacts between asperity summits of the rougher and the smoother surface. As the smoother surface attains a certain minimum value of R_{q1}, the asperity summits of the rougher surface (which are still high enough) start contacting more and more valleys of its contact mate's topography. As a result, wear dimples or grooves are expected to be produced on the smoother surface in a later stage, which should lead its roughening (i.e. to an increase in R_{q1}).

It can also be seen from Figs. 4(a) and 4(b) that the major changes with the surface topographies have occurred within the first $2 \cdot 10^5$ cycles (which is similar to "running-in" behaviour). Another interesting observation is that the roughness evolution process in this case involves gradual reduction of the high-frequency and low-amplitude roughness component, while the lower-frequency higher-amplitude component (a kind of "waviness") persists. Similar roughness behaviour has been found in the experiments.

The corresponding gradual change in the contact pressure and shear traction distributions is given in Fig. 5, top and bottom, respectively. It corresponds to the time step when the roughness sample is located at the centre of the contact. In Fig. 5, it can be seen that in the first load cycle, the pressures are limited by plasticity. As the number of cycles increases, the surface roughness is reduced (mainly due to wear), increasing the Λ value and reducing the surface tractions.

Once the model is able to predict the topography evolution of both contacting surfaces with increasing number of loading cycles, it looks useful to monitor the corresponding history of the key roughness parameter. In the present work, the time change of the roughness root-mean-square parameter R_q was examined as a function of the various operating conditions — the initial roughness combination, the lubrication quality, the load, and the slide-to-roll ratio.

In the example with combination of initial roughness, $R_{q1} = 48$ nm vs. $R_{q2} = 477$ nm, is shown in Fig. 6. One surface is rougher and the other is smoother. For its rougher contact mate, the roughness was scaled up with a factor of 2.5. The three values of the lubrication quality,

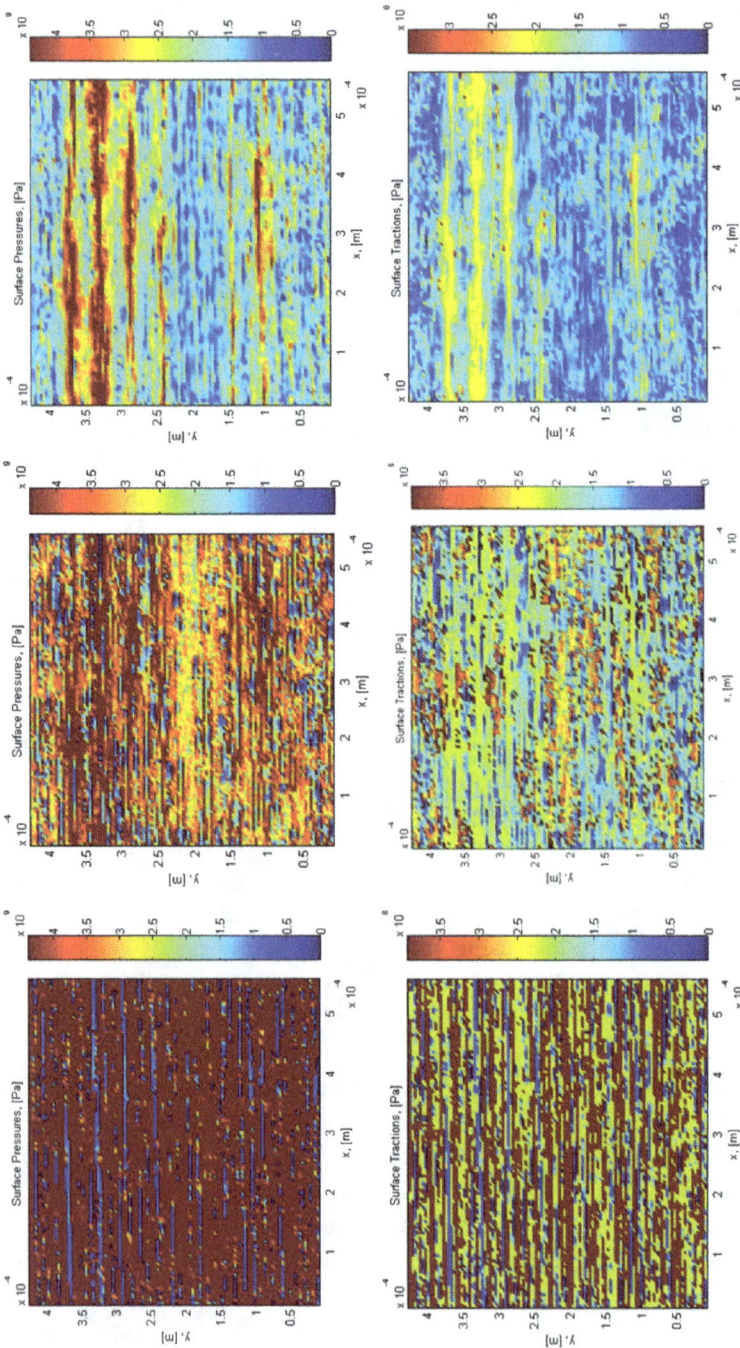

Fig. 5:   Here the corresponding distribution of pressure (top) and shear tractions (bottom) in the contact; left – initial; middle – after $N = 2 \times 10^5$ cycles; right – after $N = 10^6$ cycles. For all the results: left – initial; middle – after $N = 2 \cdot 10^5$ cycles; right – after $N = 10^6$ cycles. Figure reprinted from [30] with permission from Sage.

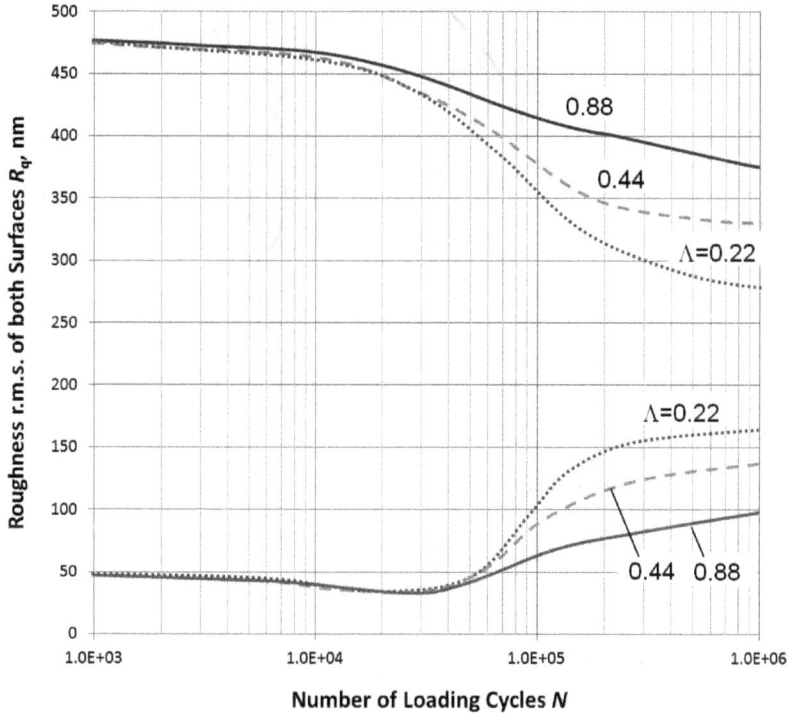

Fig. 6: The R_q values of the rougher and the smoother contacting surfaces as a function of time, as predicted by the model. Example 2: $R_{qi1} = 48$ nm, $R_{qi2} = 477$ nm, $p_0 = 1.5$ GPa, $S = 0.02$, $U = 1$ m/s, $k_{lub} = 3 \cdot 10^{-11}$, $k_{dry} = 3 \cdot 10^{-10}$. The two surfaces accumulate the loading cycles with identical rate. Figure reprinted from [30] with permission from Sage.

identical to those in the previous example, were provided by controlling the lubricant viscosity: $\eta_0 = 10.8$ mPa·s, 30 mPa·s and 83.4 mPa·s, were used in this case to obtain $\Lambda = 0.22, 0.44$, and 0.88, respectively. While the rest of the operating conditions remained unchanged (see legend of Fig. 6). Despite the obvious difference in the initial roughness combination, one can conclude from Fig. 6 that the trend of roughness behaviour is similar to the previous case: the rougher surface is gradually smoothing, while the smoother one first gets even smoother, but then it roughens later in time.

5. Wear-Fatigue Interaction in Contact Profiles

Lubricated contacts in gears and rolling bearings often suffer from abrasive wear or massive micropitting, which in time will modify the macroscopic

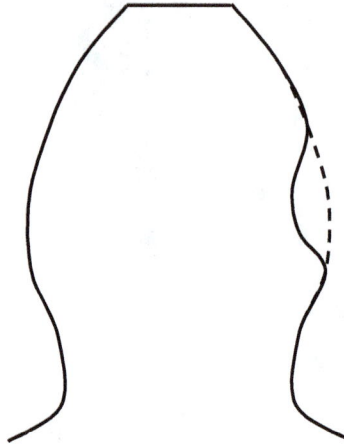

Fig. 7:　Example of modified contact profiles by wear (gear tooth involute).

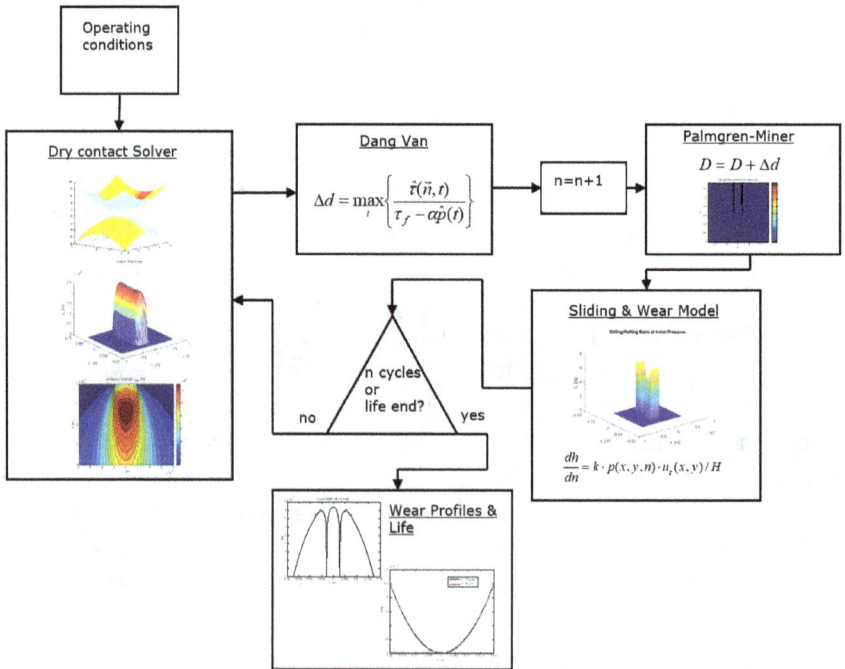

Fig. 8:　Modified Wear-Fatigue model for considerations of contact profile effects. Figure inspired from [30].

profile of the contact (see Fig. 7). In these cases the previous model can be simplified by assuming the following:

1. The micro-geometry (roughness) will have very little effect, so it can be disregarded.
2. The hydrodynamic pressures will not be too different from the dry contact pressures, so the Reynolds equation can be omitted and lubrication will only affect the friction coefficient.

With these modification the previous model reduces to the flow chart given in Fig. 8, as presented in [30]. To give an example of the use of this model, consider the case of the cylindrical thrust roller bearing (designation 81212) given in [30] see (Fig. 9). Which has a linear variation of sliding from the roller centre (zero sliding) towards the roller ends (maximum sliding). In the reference, calculations are done for this case showing the contact pressures for the first load cycle and the final load cycle (31 million) when the first failure point appears, see Fig. 10. In this figure the pressure peaks at the edges have been eliminated by considering the end-roller radii from manufacture at those locations.

For this case, an equivalent experiment was carried out in the laboratory using rollers with artificially modified profiles with the results showing similar behaviour as the model, including the location of the failure. This is shown in Fig. 11.

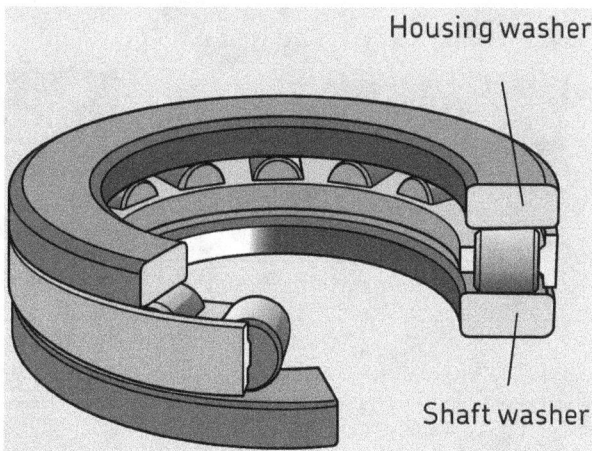

Fig. 9: Cylindrical thrust roller bearing schematics. Figure courtesy of SKF.

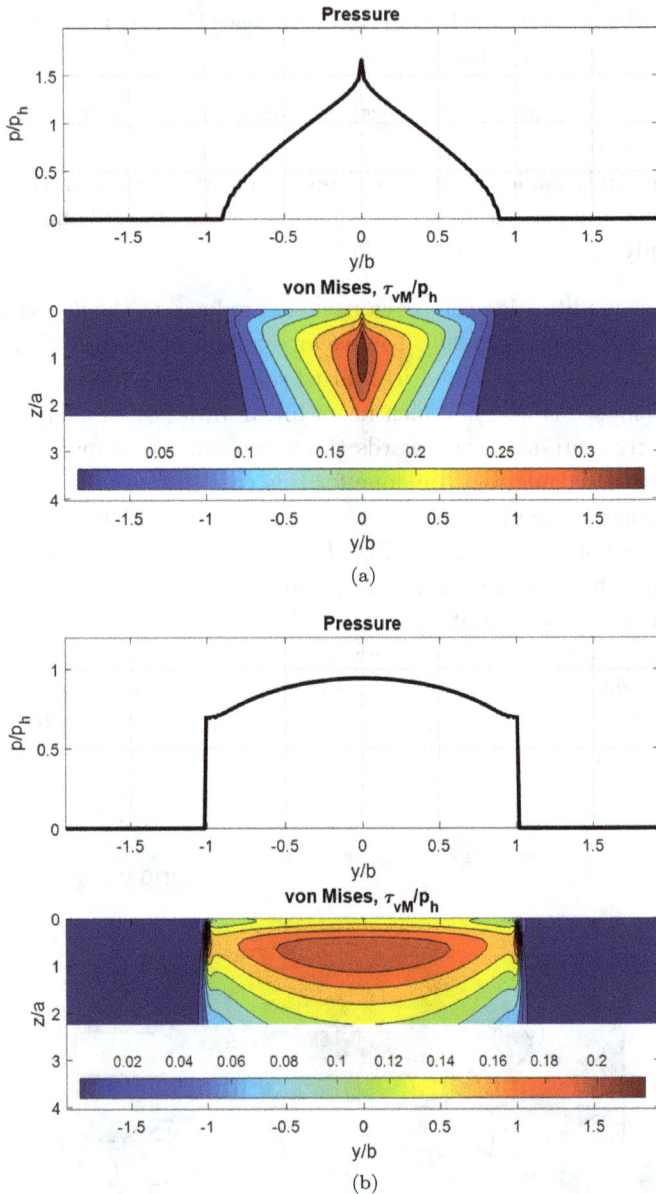

Fig. 10:   Dimensionless contact pressures and von Mises stresses calculated for the rolling bearing at the (a) initial load cycle; (b) the final (31 million) after wear when the failure is produced. Figure inspired from [30].

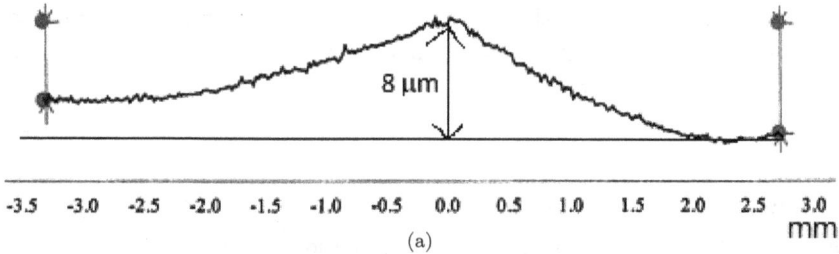

(a)

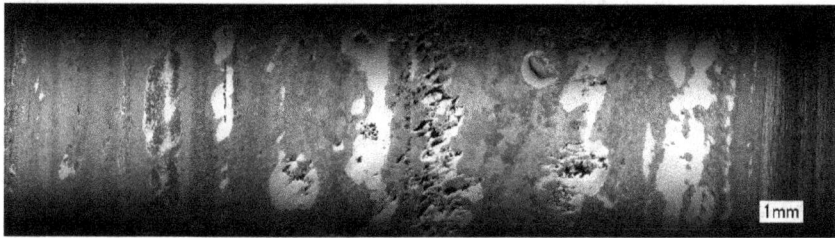

(b)

Fig. 11: Experimental results using bearings with artificially modified profiles. (a) Artificially modified profile of tested bearings; (b) damaged rolling element of a bearing after test.

6. Conclusions

A mixed-lubrication model is used in the calculation of the Archard local wear coefficient under micro-EHL conditions. This wear coefficient is used for the estimation of the 3D roughness evolution of both surfaces during cyclic loading in a lubricated rolling/sliding contact. Equally, in reference [30] experiments are carried out under similar conditions where the surface topographies of both surfaces are monitored as the number of loading cycles progresses. The results in terms of R_q values are compared with the predictions of the model finding a good agreement in behaviour and trends.

The main advantages of the present model are as follows:

- The micro-geometries of both surfaces and the associated pressures are followed-up in time as they travel within the EHL contact, during the numerous contact cycles.
- The wear coefficient is estimated locally, as a function of location and time and then averaged over a contact in area and within a contact cycle.

- The model can be greatly simplified when the interest is not the micro-geometry but the contact profile modification due to massive wear and its interaction with fatigue.

Nomenclature

a	Hertzian semi-width in the x-direction	[m]
A	Contact area	[m^2]
A_s	Sliding area in a certain time	[m^2]
b	Hertzian semi-width in the y-direction	[m]
B	Bulk modulus of the lubricant, $d\rho/dp = \rho/B$	[Pa]
E'	Combined elasticity modulus for EHL	[Pa]
F	Normal load in the contact	[N]
Δh	Wear thickness at a particular point and time during a load cycle, $\Delta h = h_w/n_t$	[m]
h	Wear removed surface layer thickness in a certain time	[m]
h_c	Central film thickness	[m]
h_w	Average thickness of wear removal layer per load cycle	[m]
H	Hardness of the steel	[Pa]
k	Dimensionless Archard wear coefficient	[-]
k_p	Dimensional wear coefficient, $k_p = kt/N$	[s]
L_x, L_y	Length of the roughness sample in the x, y directions	[m]
p	Contact pressure	[Pa]
p_0	Maximum Hertzian pressure	[Pa]
$\bar{p}$	Mean pressure in the analysed roughness window in the model	[Pa]
l	Contact length in the rolling direction (x)	[m]
N	Number of load cycles	[-]
n_t	Number of time steps for roughness movement in a load cycle	[-]
r	Roughness sinusoidal component	[m]
R_x	Radius in rolling direction	[m]
R_y	Radius in transverse direction	[m]
R_q	r.m.s. for roughness	[m]
R_{qi}	r.m.s. for roughness, initial	[m]
s	Sliding distance	[m]
S	Sliding/rolling ratio, $S = (u_2 - u_1)/\bar{u}$	[-]
t	Time, time of passage of both sliding surfaces throughout the contact zone with the sliding speed	[s]

T	Temperature	[°C]
Δt	Time step in the roughness movement within a load cycle	[s]
u	Surface speed	[m/s]
$\bar{u}$	Mean speed, $\bar{u} = (u_2 + u_1)/2$	[m/s]
u_s	Sliding speed, $u_s = (u_2 - u_1)$	[m/s]
v	Elastic displacement sinusoidal component	[m]
V	Wear volume	[m³]
x	Rolling direction co-ordinate	[m]
y	Transverse to rolling co-ordinate	[m]
z	Normal to surface co-ordinate	[m]

Greek Symbols

α	Viscosity-pressure coefficient of the lubricant	[Pa^{-1}]
δ	Indicates wave fluctuation	[-]
λ_x, λ_y	Wavelengths in the x and y directions	[m]
Λ	Lubrication quality parameter for film thickness, $\Lambda = h_c/R_q$	[-]
η_0	Lubricant viscosity at ambient pressure	[Pas]
η_x, η_y	Lubricant equivalent viscosities in the x and y directions for a non-Newtonian fluid	[Pa]
ω_x, ω_y	Wavelengths in the x and y directions	[1/m]
ρ	Density sinusoidal component	[kg/m³]
τ_0	Eyring stress of the lubricant	[Pa]
τ_m	Mean shear stress	[Pa]

Subscripts

1,2	Corresponds to body 1 or 2, respectively	[-]
a	Indicates amplitude of fluctuation	[-]
lub	Corresponds to fully lubricated patches of the contact area	[-]
dry	Corresponds to boundary lubricated patches of the contact area	[-]

References

[1] P. M. Lugt and G. E. Morales-Espejel, A review of elasto-hydrodynamic theory, *Tribol Trans.* **54**, 470–496 (2011).
[2] R. Gohar, *Elastohydrodynamics*, 2nd edition. Imperial College Press, UK, 2001.

[3] G. E. Morales-Espejel and A. Gabelli, Rolling bearing seizure and sliding effects on fatigue life, *Proc IMechE, Part J, J Eng Tribol.* **233**(2), 339–354 (2019).

[4] S. Wu and H. S. Cheng, A sliding wear model for partial EHL contacts, *Trans. ASME, J. of Tribology*, Vol. 113, pp. 134–141, 1991.

[5] G. K. Nikas, R. S. Sayles, and E. Ioannides, Effects of debris particles in sliding/rolling elastohydrodynamic contacts, *Proc IMechE, Part J, J Eng Tribol.* **212**, 333–343 (1998).

[6] D. Savio, N. Fillot, P. Vergne, H. Hetzler, W. Seemann, and G. E. Morales-Espejel, A multiscale study on the wall slip effect in a ceramic–steel contact with nanometer-thick lubricant film by a nano-to-elastohydrodynamic lubrication approach, ASME, *J Tribol.* **137**, 031502-1–13 (2015).

[7] A. Porras-Vazquez, L. Martinie, P. Vergne, and N. Fillot, Independence between friction and velocity distribution in fluids subjected to severe shearing and confinement, *Phys Chem Chem Phys.* **20**, 27280–27293 (2018).

[8] T. E. Tallian, Y. P. Chiu, D. F. Huttenlocher, J. A. Kamenshine, L. B. Sibley, and L. E. Sindlinger, Lubricant films in rolling contact of rough surfaces, *ASLE Trans.* **7**, 109–126 (1964).

[9] T. E. Tallian, The theory of partial elastohydrodynamic contacts, *Wear.* **21**, 49–101 (1972).

[10] K. L. Johnson, J. A. Greenwood, and S. Y. Poon, A simple theory of asperity contact in elastohydrodynamic lubrication, *Wear.* **19**, 91–108 (1972).

[11] J. A. Greenwood and J. B. P. Williamson, Contact of nominally flat surfaces, *Proc. R. Soc. (London)* **295A**, 300–319 (1966).

[12] E. R. M. Gelinck and D. J. Schipper, Calculations of stribeck curves for line contacts, in *Proceedings of the 5th International Tribology Conference, Australia*, 1998, pp. 75–80.

[13] E. R. M. Gelinck, *Mixed Lubrication of Line Contacts*, PhD Dissertation, University of Twente, The Netherlands, 1999.

[14] X. Jiang, D. Y. Hua, H. S. Cheng, X. Ai, and S. C. Lee, A mixed elastohydrodynamic lubrication model with asperity contact, ASME, *J Tribol.* **121**, 481–491 (1999).

[15] Y. Z. Hu, and D. Zhu, A full numerical solution to the mixed lubrication in point contacts, ASME, *J Tribol.* **122**, pp. 1–9 (2000).

[16] A. C. Redlich, D. Bartel, and L. Deters, Calculation of EHL contacts in mixed-lubrication regime, in D. Dowson *et al.* (eds.), *Tribological Research and Design for Engineering Systems*. Elsevier B.V., Amsterdam, The Netherlands, 2003, pp. 537–547.

[17] J. Q. Wang, D. Zhu, H. S. Cheng, T. Yu, X. Jiang, and S. Liu, Mixed lubrication analysis by macro–micro approach and a full-scale EHL model, ASME, *J Tribol.* **126**, 81–91 (2004).

[18] N. Deolalikar, F. Sadeghi, and S. Marble, Numerical modelling of mixed lubrication and flash temperature in EHL elliptical contacts, ASME, *J Tribol.* **130**, 011004–011023 (2008).

[19] G. E. Morales-Espejel, A. W. Wemekamp, and A. Félix-Quiñonez, Micro-geometry effects on the sliding friction transition in elastohydrodynamic lubrication, *Proc IMechE, Part J, J Eng Tribol.* **224**, 621–637 (2010).

[20] H. M. Stanley and T. Kato, A FFT-based method for rough surface contact. ASME, *J Tribol.* **119**, 481–485 (1997).

[21] G. E. Morales-Espejel, C. H. Venner, and J. A. Greenwood, Kinematics of Transverse real roughness in elastohydrodynamically lubricated line contacts using fourier analysis, *Proc IMechE, Part J, J Eng Tribol.* **214**, 523–534 (2000).

[22] C. J. Hooke, Roughness in rolling–sliding elastohydrodynamic lubricated contacts, *Proc IMechE, Part J, J Eng Tribol.* **220**, 259–271 (2006).

[23] C. J. Hooke, K. Y. Li, and G. E. Morales-Espejel, Rapid calculation of the pressures and clearances in rough, rolling–sliding elastohydrodynami-cally lubricated contacts. Part 1: Low-amplitude, sinusoidal roughness, *Proc IMechE, Part C, J Mech Eng Sci.* **221**, 535–550 (2007).

[24] C. J. Hooke, K. Y. Li, and G. E. Morales-Espejel, Rapid calculation of the pressures and clearances in rough, rolling–sliding elastohydrodynamically lubricated contacts. part 2: general, non-sinusoidal roughness, *Proc. IMechE, Part C, J Mech Eng Sci.* **221**, 551–564 (2007).

[25] J. J. Kalker, Variational principles in contact elastostatics, *J Inst Math Appl.* **20**, 199–219 (1997).

[26] C. H. Venner and A. A. Lubrecht, Multigrid techniques: a fast and efficient method for the numerical simulation of elastohydrodynamically lubricated point contact problems, *Proc. IMechE, Part J, J Eng Tribol.* **214**, 43–62 (2000).

[27] G. E. Morales-Espejel, P. M. Lugt, J. van Kuilenburg, and J. H. Tripp, Effects of surface micro-geometry on the pressures and internal stresses of pure rolling EHL Contacts, *Tribol Trans.* **46**(2), 260–272 (2003).

[28] G. E. Morales-Espejel, Surface roughness effects in elastohydrodynamic lubrication: A review with contributions, *Proc. IMechE, Part J, J Eng Tribol.* **228**(11), 1217–1242 (2014).

[29] C. H. Hooke, The behaviour of low-amplitude surface roughness under line contacts: non-newtonian fluids, *Proc. IMechE, Part J, J Eng Tribol.* **214**, 253–265 (2000).

[30] G. E. Morales-Espejel, V. Brizmer, and E. Piras, Roughness evolution in mixed lubrication condition due to mild wear, *Proc IMechE, Part J, J Eng Tribol.* **229**(11), 1330–1346 (2015).

[31] J. F. Archard, Contact and rubbing of flat surface, *J Appl Phys.* **24**(8), 981–988 (1953).

[32] G. E. Morales-Espejel and V. Brizmer, Micropitting modelling in rolling-sliding contacts: Application to rolling bearings, *Tribol Trans.* **54**, 625–643 (2011).

Chapter 5

In-silico Analytical Wear Predictions: Application to Joint Prostheses

Francesca Di Puccio* and Lorenza Mattei

*Department of Civil and Industrial Engineering,
Università di Pisa, Largo Lucio Lazzarino 2, 56126, Pisa, Italy
dipuccio@ing.unipi.it

Wear predictions are an attractive alternative to expensive and long experimental tests. This chapter presents the basic elements for developing a predictive wear model by means of an analytical approach, together with examples of application to artificial joints. With respect to Finite Element simulations, the analytical approach is a good option for achieving rapid indications, e.g. for comparing design variables. The general procedure for wear predictions includes kinematic and contact analyses, wear law definition, wear estimations of both local and global quantities, i.e. wear depth and volume. Archard's wear law is mainly considered for abrasive–adhesive wear modelling, but other laws for describing the anisotropic wear behaviour of the plastic materials are also presented.

The application to hip and shoulder implants enabled sensitivity analyses of wear quantities to geometrical features and loading conditions. Results were in good agreement with literature studies, supporting the use of this simple approach.

1. Introduction

Wear is a tribological phenomenon occurring on rubbing surfaces, where contact actions and relative sliding cause a loss of volume with surface damage. Therefore, wear is considered one of the main causes of failure of mechanical components, from gears to joint replacements (Fig. 1). Wear resistance is generally determined with expensive and time-consuming

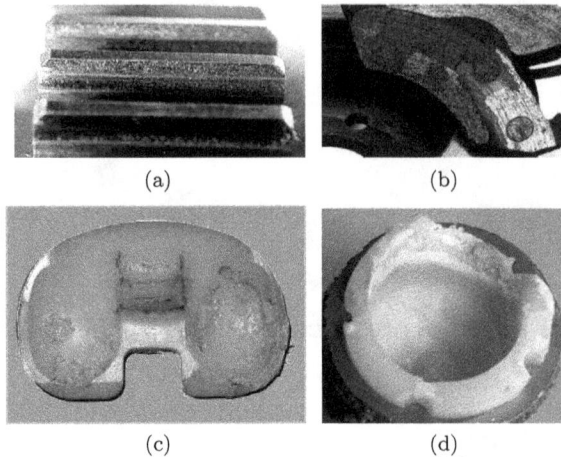

Fig. 1: Examples of worn surfaces of mechanical components and biomedical devices: gear (a), disc brake (b), tibial plate of a total knee implant (c), acetabular cup of hip implants (d).

experimental tests under simplified laboratory conditions. Thus, *in-silico* wear predictions represent an attractive alternative to experimental campaigns, being cheaper and more rapid. Moreover, they can provide indications on wear behaviour of rubbing surfaces in realistic working conditions. However, wear simulations are not so widespread today, mainly because the development of models requires some peculiar attentions and it is not easy to check the reliability of results.

This chapter aims to describe *in-silico* wear predictions by means of an analytical approach, in order to provide some practical indications to researchers entering this field. As examples, applications to investigate wear in hip and shoulder prostheses are described. An alternative *in-silico* approach based on the Finite Element method is presented in Chapter 6.

2. Wear Modelling: Overview

The general procedure for wear estimation is briefly outlined in this section. It consists of several steps:

(a) Definition of the relative pose of the two bodies in contact (for the whole simulated time interval) and identification of the nominal contact point. This pose can be determined by solving geometrical (pin-on-disc, cams, etc.) or equilibrium conditions (e.g. hip implants). The latter case is usually more complicated.

(b) Mapping of the sliding velocity of the articulating surfaces.

(c) Evaluation of the contact actions between the surfaces.

(d) Estimation of wear according to the adopted wear law. Both local wear depth h and the total volume loss can be defined as outputs of this procedure.

Since wear modifies the geometry by removing material, a fifth step can be introduced.

(e) Geometry update (at a proper simulation time).

The procedure is then repeated cyclically until the whole wear process is simulated.

The main difference between the analytical and the FE approach is in the solution of the contact problem, step c. While FE packages can provide the solution for a general contact problem, contact pressure can be analytically estimated only in simple cases. For example, Hertz formulas can be used for non-conformal contacts, while approximated elastic foundation solutions, for extended contacts [1, 2]. Though the implementation of the geometry update is feasible also in analytical models, it is only rarely performed since the modified geometry could be unsuitable for closed form solutions for contact pressure.

The application of the described procedure is illustrated ahead in this chapter focusing on wear predictions for artificial hip and shoulder joints. First, some general observations on the wear law are reported, to highlight the basic concepts of wear formulations.

3. Wear Law

The fundamental "ingredient" for a wear predictive model is the wear law that is an analytical relationship trying to describe the wear phenomenon, usually under some simplifying assumptions. Indeed, the actual phenomenon is very complex and depends on many factors: micro and macro geometry of the contact coupling, materials, loading conditions, lubrication, temperature, etc. Their combination can result in different wear mechanisms: abrasive and/or adhesive wear, fretting, tribocorrosion [3]. Despite the large variety of wear mechanisms, most simulations are based on Archard's law for abrasive–adhesive wear. It states that, for a translating body, the volume loss V is proportional, via the wear coefficient k, to the product of the normal contact force L_N and the sliding distance s

$$V = kL_Ns, \tag{1}$$

or, alternatively

$$V = \frac{K}{H} L_N s, \tag{2}$$

where H represents the material hardness. The main advantage of the Archard law lies in its simplicity and in the good correlation with experimental evidence.

3.1. *Unilateral or bilateral wear*

In Eq. (1), the total volume loss V can come from one or both the bodies in contact. When one part is much softer than the other, unilateral wear can be assumed, i.e. only one body gets worn, as in metal on plastic contacts. In case of bilateral wear, both parts get worn and two equations as Eq. (1) have to be written to differentiate the wear behaviour of the two bodies, i.e.

$$V_1 = k_1 L_N s, \tag{3}$$

$$V_2 = k_2 L_N s, \tag{4}$$

with $V = V_1 + V_2$ and $k = k_1 + k_2$. However, distinct values of k are only rarely reported in the literature, as well as indications of the separate volumes [4–6]. Most frequently, only one value of k is provided and/or used even when wear is not equally distributed between the bodies [7, 8]. This may be due to the fact that wear, both as phenomenon and as coefficient k, is strongly connected with the coupling more than with single parts. In order to differentiate wear between the two bodies, the present authors have proposed a dimensionless parameter α, named wear partition factor [5], so that

$$k_1 = \alpha k, \quad k_2 = (1 - \alpha)k, \tag{5}$$

and showed its use in simulations in [5, 9].

3.2. *Local formulation of Archard's wear law*

As previously stated, Eq. (1) holds only when the relative motion between the rubbing surfaces is a simple translation. In order to deal with a more general case or Archard's wear law should be written in a local form to evaluate the wear depth h at a point P moving on a trajectory γ under a pressure p

$$h(P) = k \int_\gamma p(P, s) \, ds, \tag{6}$$

where s is the arc length along γ. Additionally, the relation $s = s(t)$ can be introduced in Eq. (6) and h derived to obtain the local wear rate

$$\dot{h}(P,t) = k \, p(P,t)|\boldsymbol{v}_s(P,t)|, \tag{7}$$

where $\boldsymbol{v}_s$ is the sliding velocity.

A further generalisation of the Archard law can also be found as

$$\dot{h}(P,t) = k \, p(P,t)^m |\boldsymbol{v}_s(P,t)|^n, \tag{8}$$

where pressure and sliding velocity are assumed to have the exponents m and n, respectively.

These local and time-dependent formulations are usually preferred in numerical implementations of wear, in continuous or discrete version.

The total wear volume can be obtained by integrating h over the contact/worn area at a given time/number of cycles.

$$V(t) = \int_A h(P,t) \, dA. \tag{9}$$

However, since the geometrical up-date is not implemented, the wear after n loading/wear cycles can be obtained simply by multiplying h and V estimated for a single cycle by n.

3.3. *Experimental evaluation of the wear coefficient/law*

The key element in the wear law is the wear coefficient, i.e. k in Eq. (1) or $k_{1,2}$ in Eqs. (3–4). Usually, it is obtained from standard tests, as pin-on-disc or pin-on-plate wear tests, where a constant normal load L_N is applied to a pin in steady state or cyclic motion with respect to its counterpart. The volume loss V^{exp} is measured by weighting the wear debris or directly the test samples at selected time intervals and/or at the end of the test. Separate measures of samples are needed to estimate $k_{1,2}$. The wear coefficient can be estimated by comparing experimental and numerical wear volumes (V from Eq. (1)) obtained under the same conditions, at a given sliding distance s, thus considering $V = V^{exp}$ whence

$$k = V^{exp}/(L_N \, s). \tag{10}$$

However, the estimation of a single value for k implies a steady-state trend for the phenomenon that in general is not realistic. In most cases, two phases can be identified: a first period of running-in when wear is more rapid and a second phase where wear proceeds more slowly Fig. 2. This behaviour is described by (at least) two distinct values for k.

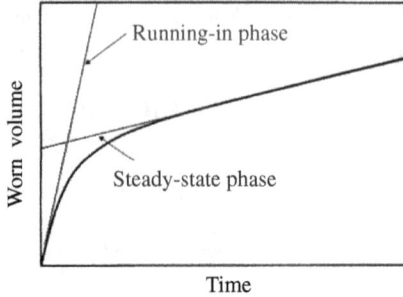

Fig. 2: Typical two-phases trend of the wear volume during wear damage process: (i) initial running-in phase at high wear rate, and (ii) steady-state phase at low wear rate.

Another crucial issue is the correlation between results of standard pin-on-disc and pin-on-plate wear tests and the behaviour of the components in operative conditions. Since wear depends on many factors, laboratory tests should try to replicate as close as possible the operative conditions of the rubbing parts under examination: geometry, materials, surface finishing, temperature, pressure, lubrication and so on. However, this is almost infeasible and caution should always be taken when using data from tests in practical applications.

3.4. *Anisotropic wear: Cross-shear phenomenon*

When dealing with wear of plastics, as ultra high molecular weight polyethylene (UHMWPE), the wear coefficient of Archard's wear law is frequently modified to characterise the anisotropic material behaviour. In fact, experimental observations prove that when UHMWPE is subjected to multidirectional sliding against a hard counterpart, a principal molecular orientation (PMO) is set in the polymeric chains [10, 11]. In the PMO direction, the wear resistance increases, while in the direction perpendicular to the PMO, named the Cross-Shear direction, wear can occur more easily, Fig. 3.

This anisotropic behaviour depends on the point of the surface through a scalar quantity known as the cross-shear ratio (CS) defined as the ratio between the frictional work done perpendicularly to the PMO ($W_\perp$) and the total frictional work (W_tot)

$$\text{CS}(P) = \frac{W_\perp}{W_\text{tot}} = \frac{\int_0^T fp(P,t)|\boldsymbol{v}(P,t)|\sin^2(\zeta(P,t))dt}{\int_0^T fp(P,t)|\boldsymbol{v}(P,t)|dt}, \tag{11}$$

Multidirectional sliding   Molecular orientation

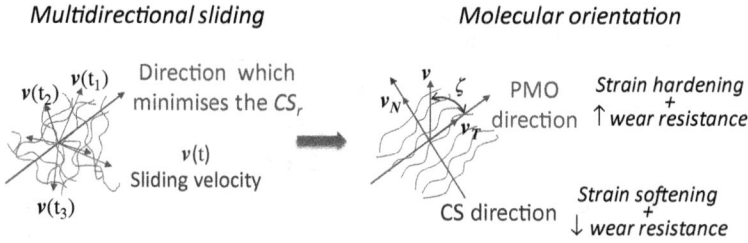

Fig. 3: Multi-directional sliding and molecular orientation with identification of the PMO and CS directions. (Adapted and reprinted from [12] with permission from Elsevier.)

where ζ is the angle between the PMO direction and the sliding velocity (Fig. 3). The wear coefficient in Eq. (1) becomes a function of the point as $k\,(\mathrm{CS}\,(P))$. The main difficulty in introducing this anisotropy in the model is that the PMO direction is unknown, and it comes from the solution of an optimization problem where PMO is the direction which minimises $\mathrm{CS}(P)$ or, alternatively, $W_\perp$.

Despite the above definition, a kinematic cross-shear ratio CS_k is used more frequently

$$\mathrm{CS}_k\,(P) = \frac{\int_0^T |\boldsymbol{v}\,(P,t)|\sin^2\,(\zeta\,(P,t))\,dt}{\int_0^T |\boldsymbol{v}\,(P,t)|\,dt}. \tag{12}$$

It comes in case of constant coefficient of friction (CoF) and when the current pressure value is replaced with its average value $\bar{p}$ both in the numerator and denominator so that it can be simplified (more details in [12]).

The dependence of k on CS_k and on both CS_k and $\bar{p}$ have been proposed by Kang *et al.* [13, 14] for hip implants and are reported in Table 1, with other formulations of k that will be used in the next section. A completely different law can also be considered, proposed by Liu *et al.* [15], where the nominal contact area A replaces the load

$$V = k_c\,As. \tag{13}$$

This synthetic overview of the wear laws does not pretend to provide an exhaustive discussion of the matter, but to present the key issues when developing or choosing a wear law for a predictive model. Other types of laws can be found in the literature, for some specific coupling of materials.

Table 1:   Wear laws and wear coefficient expressions proposed in the literature for metal-on-UHMWPE couplings, discussed in [12].

Wear law	k/k_c	k/k_c expression $(mm^3/(N\ m))$	Ref.
$k\,L_N s$	k cost	$1.066\ 10^{-6}$	[16, 17]
	$k(R_a)$	$(8.68\,R_a + 1.51)\ 10^{-6}$	[18]
	$k(CS_k)$	$(0.328\ln(CS_k) + 1.62)\ 10^{-6}$	[13]
	$k(CS_k, \bar{p})$	$exp(-13.1 + 0.19\ln(CS_k) - 0.29\bar{p})\ 10^{-6}$	[14]
$k_c\,As$	$k_c(CS_k)$	$\begin{cases} 32\,CS_k + 0.3)10^{-9}CS_k \leq 0.04 \\ 1.9\,CS_k + 1.6)10^{-9}0.04 < CS_k \leq 0.5 \end{cases}$	[15]

4. Prediction of Wear in Artificial Joints

Predictive wear models of artificial joints are particularly important for improving their design, comparing materials, and above all, for simulating realistic working conditions, similar to those observed *in-vivo*. The current standards for the experimental evaluation of wear consider fairly simple loading and kinematics conditions (e.g. a kind of simplified gait for hip [30] and knee [31] implants), generally implemented in special simulators. Moreover, experimental campaigns are very expensive and long, requiring about $2-5\,10^6$ test cycles. *In-silico* models can provide indications in shorter times, with a remarkable economic advantage. Above all, they can simulate more closely what happens *in-vivo*, during the execution of several different motor tasks. One point of interest in this research is to understand to what extent the regulations are able to evaluate the actual behaviour of the prostheses and eventually improve the tests.

For this purpose, predictive wear models were developed based on simple analytical relationships. Indeed, analytical models are rather simple and efficient but do not take into account the modification of the geometry with wear, so they are suitable for "small wear" and a first estimate. FEM models, presented in Chapter 6 allow for geometry updates but are quite expensive from a computational point of view [5, 19]. In case of joint replacement, where PE is used in different forms as UHMWPE, XLPE, etc., analytical models can be convenient for including the CS effect.

4.1. *Hip and shoulder implants*

As example of the analytical approach, the models for two artificial joints are described: those for the widespread hip arthroplasties (HA) and reverse

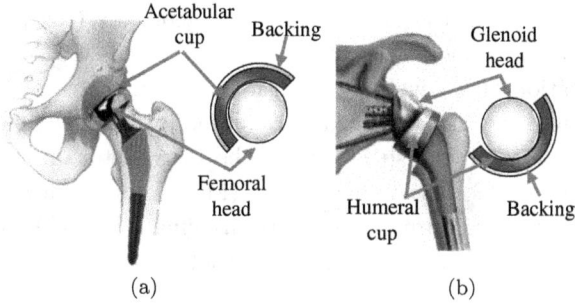

Fig. 4:   Hip implant (a) and reverse total shoulder implant (b): images (on the left) and equivalent ball-in-socket geometry (on the right).

total shoulder arthroplasties (RTSA), re-evaluated in the last decades thank to the modern Grammont design [20] (Fig. 4). Though HAs and RTSAs aim at restoring the biomechanics of quite different human synovial joints and, in particular, experience different loading and kinematic conditions, these implants have some common design characteristics and consequently can be treated similarly in wear modelling. Indeed, both of them have a ball-in-socket geometry with the femoral/glenoid head coupled with the acetabular/humeral cup (Fig. 4) and are made of the same materials, i.e. plastic/metal for the cup and metal/ceramic both for cup and head [21, 22].

Our previous investigations were focused on:

1. Metal-on-plastic (MoP) HAs [12];
2. Metal-on-metal (MoM) HAs [9, 23–25];
3. Metal-on-plastic RTSAs [26, 27].

The analytical approach combined with a parametric formulation of the model allowed to easily modify model input, from the geometry to the loading/kinematics curves, and the wear laws/coefficients. For this purpose, the software Mathematica® or Mathcad® were employed as they offer powerful tools for symbolic calculus. The developed models were applied for addressing some critical issues highlighted by the literature review and underpinned as follows:

(a) Modelling the cross-shear phenomenon in MoP implants and comparison of the wear laws adopted for UHMWPE [12, 26].
(b) Investigations on the effect of the loading and kinematic conditions on wear, by comparing wear predicted for *in-vivo* and *in-vitro* joint simulator BCs [12, 24].

(c) Investigations on the effect of implant geometry (size and clearance, related to the dimensional tolerance) on wear [24, 26].

(d) Presentation of a novel approach for estimating distinct wear factors for the implant surfaces in case of bilateral wear, i.e. MoM HAs [9].

(e) Investigations on the specificity of the wear factor on the tribological scenario: effect of the implant type (MoP HAs vs MoP RTSAs [26]), the implant structure (total vs resurfacing MoM HAs [9]), and the boundary conditions [9] on the wear coefficient.

4.2. Models description

A general analytical wear model of a ball-in-socket couple in case of bilateral wear is presented in this section. The wear models of HAs and RTSAs can be obtained from the general one, making specific simplifications for the implant type.

4.2.1. Model assumptions

All wear models were based on two hypotheses largely adopted in the literature:

1. The surfaces are considered in dry contact, the lubrication effect being taken into account in the wear coefficient empirically derived.

2. Adhesion and abrasion are considered as the main wear mechanisms, neglecting all the other ones occurring in *in-vivo* (e.g. fatigue, corrosive wear).

Further common basic assumptions are:

3. Geometrical variations due to wear do not affect contact mechanics. That implies that the wear model is suitable for wear prediction during the running-in, i.e. the highest wear rate phase.

4. The contribution of the elastic deformation of cup/head to sliding velocity can be disregarded with respect to the rigid body movements.

Some specific assumptions were called when modelling MoP implants ($'$):

5$'$. The wear of the metallic component (head) is negligible with respect to the wear of the plastic side (cup).

6$'$. The contact is considered frictionless, being $f < 0.06$.

7$'$. Creep effects are negligible.

or MoM implants ($''$):

5$''$. The friction coefficient is constant during the running-in phase ($f = 0.2$).

6$''$. Contact pressure distribution is not affected by friction, i.e. can be estimated from frictionless contacts.

7$''$. The wear coefficient is constant.

It is worth emphasising that, according to these assumptions, the wear is modelled as unilateral in MoP implants, while as bilateral in MoM ones.

4.2.2. *Wear laws/model*

In these applications, contact conditions vary in space and time, therefore the local and instantaneous form of the wear law is preferred. More specifically, Eq. (8) is used for describing the linear wear rate of a generic point on the cup and head surface, labelled P_c and P_h, respectively. Similarly, the cup/head wear coefficients are denoted as k_c/k_h.

4.2.3. *Input data*

The model inputs are listed as follows:

- Geometry: The geometry of ball-in-socket implants is described by the head and cup radii, r_h and r_c, respectively, and by the liner thickness t_c (Fig. 5(a)). The position of the cup in the pelvic/humeral bone is specified by the anteversion (α) and the inclination (β) angles (Fig. 5(b)). The models were formulated for a left implant using a fixed global frame

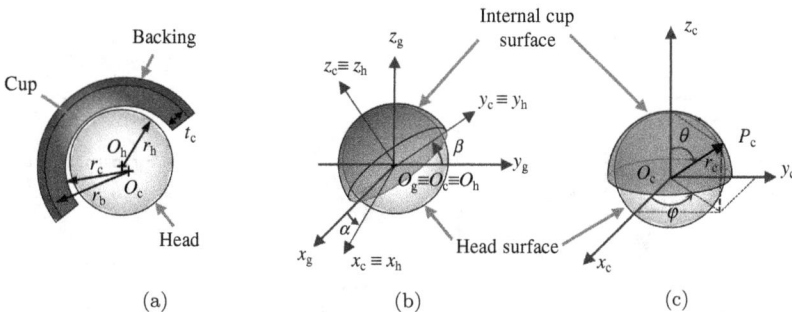

Fig. 5: Geometry (a) and coordinate frames (b) of wear models of HA and RTSA. The global frame and the local frames of head and cup are represented in the reference configuration with no loading and null rotations (b). Spherical coordinates used in model implementation (c). (Adapted and reprinted from [12] with permission from Elsevier.)

$C_g = \{O_g; x_g, y_g, z_g\}$ and two mobile frames fixed respectively on the cup, $C_c = \{O_c; x_c, y_c, z_c\}$ and on the head, $C_h = \{O_h; x_h, y_h, z_h\}$, where O_c/O_h are head/cup centres (Fig. 5(b)).

- Materials: Cup and head material elastic modulus $E_{h,c}$ and Poisson ratio $\nu_{h,c}$. According to the model assumptions, for MoP $k_h = 0$ while k_c is defined according to Table 1, for MoM HAs, k_h and k_c are considered constant in space and time, i.e. $k_h(P_h, t) = k_h$.

- Loading and kinematic boundary conditions (BC) are typically the gait cycle for HA and the mug-to-mouth task for RTSAs, considered the most significant for the joints. The boundary time-history of the contact force (three components) and the relative orientation between the bodies (expressed as three rotations with a specific sequence depending on data source) over a task cycle. Depending on the simulated data, for instance joint simulator or *in-vivo* BCs, the load can be applied to the head or cup, and rotations to a single or both components.

4.2.4. *Model implementation*

The model implementation consists of three main steps, as follows: (a) contact analysis; (b) kinematic analysis; (c) wear assessment. A flow chart of the model implementation is given in Fig 6.

4.2.4.1. Contact analysis

The contact analysis consists of two steps:

(a) Identification of the position of the nominal contact point K.
(b) Estimation of the contact pressure distribution.

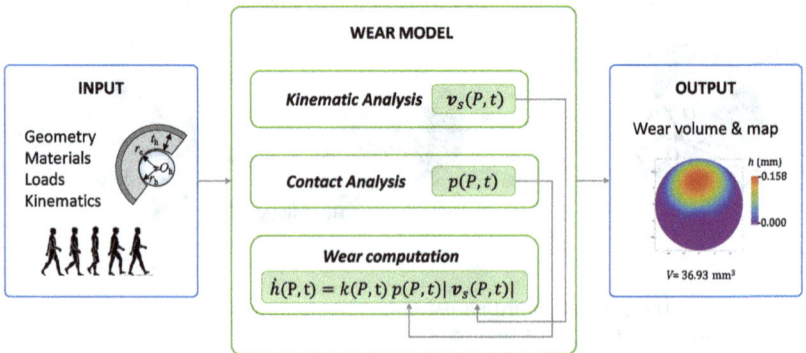

Fig. 6: Flow chart of wear model implementation.

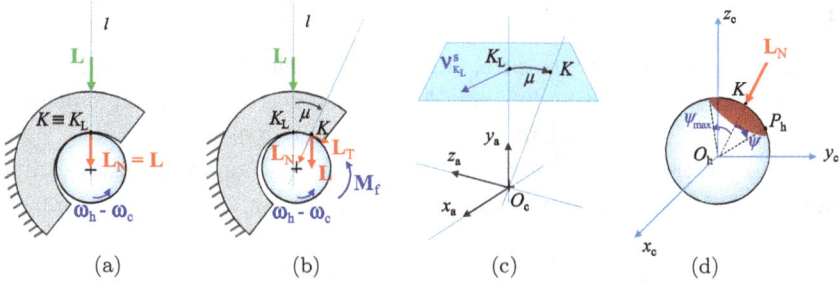

Fig. 7: Contact analysis for frictionless (a) and friction (b) contact: external load **L**, nominal contact point K. The friction causes the shift of K in the opposite direction to the sliding velocity (c) and requires a moment $\mathbf{M}_f$ to overcome frictional resistance. Example of the contact area in red (d). (Reprinted from [24] with permission from ASME.)

For a given set of BCs, the position of K is affected by the presence of the friction. In case of frictionless contact, K is instantaneously aligned with O_c and O_h, along the load line of action l (Fig. 7(a)). In the presence of friction, K is shifted from l, by an arc related to the friction angle $\mu = arctan(f)$ with f the sliding CoF in the direction opposite to the local sliding velocity, as shown in Fig. 7(b) and 7(c) [24].

The contact pressure is determined by the normal component L_N of the contact force L: in case of frictionless contact, $L_N = L$, while $L_N = L \cos \mu$ in presence of friction. The pressure distribution over the contact surfaces (red area in Fig. 7(d)) was obtained using different approaches, both analytical and numerical, depending on the implant type/materials.

In wear models of an MoP HA [12], an FE contact model was developed for estimating contact pressures under different load levels. By fitting FE results, the pressure distribution was described as

$$p(\psi) = p_{\max} \left(1 - \left(\frac{\sin(\psi)}{\sin(\psi_{\max})} \right)^2 \right)^c, \tag{14}$$

were the maximum contact pressure $p_{\max}$, the semi-angular contact width $\psi_{\max}$ (Fig. 7(d)) and c are functions of the normal load [12].

For MoM HAs, the Hertzian theory was applied though using revised reduced equivalent elastic modulus, estimated considering literature FE results [24]. Differently, Bartel's approximated formulas were assumed to estimate contact pressures in RTSAs' wear model [26].

4.2.4.2. Kinematic analysis

In addition to contact pressure, the other element of the wear law is the sliding velocity. A rigid-body kinematic analysis was therefore implemented, using rotation matrices $\mathbf{R}$ to define the pose of cup/head.

The positions of points P_c/P_h in their own reference frames are time independent and can be written for example using spherical coordinates (Fig. 5(c)) as

$$[P_i(\theta, \varphi)]_i = [O_i P_i(\theta, \varphi)]_i = r_i \begin{bmatrix} \sin\theta_i \cos\varphi \\ \sin\theta \sin\varphi \\ \cos\theta \end{bmatrix} \quad i = \text{c, h,} \qquad (15)$$

with θ_i and φ_i polar and azimuthal angles. In the following, for simplicity, the dependence on such angles will be omitted. Equation (15) means that the coordinates of P_i in frame i are equal to the components of the vector $O_i P_i$ in the same frame. Passing in the fixed frame, the coordinates change according to the following

$$[O_g P_i(t)]_g = [O_g O_i(t)]_g + \mathbf{R}_{gi}(t)[O_i P_i]_i \quad \text{with } i = \text{c,h,} \qquad (16)$$

where $\mathbf{R}_{gc}$ and $\mathbf{R}_{gh}$ are the rotation matrices that orient respectively the cup and the head frames with respect to the global fix frame. Their general form can be expressed as the scalar products between unit vectors of the reference frames, i.e.

$$\mathbf{R}_{gc} = \begin{bmatrix} \boldsymbol{i}_c \cdot \boldsymbol{i}_g & \boldsymbol{j}_c \cdot \boldsymbol{i}_g & \boldsymbol{k}_c \cdot \boldsymbol{i}_g \\ \boldsymbol{i}_c \cdot \boldsymbol{j}_g & \boldsymbol{j}_c \cdot \boldsymbol{j}_g & \boldsymbol{k}_c \cdot \boldsymbol{j}_g \\ \boldsymbol{i}_c \cdot \boldsymbol{k}_g & \boldsymbol{j}_c \cdot \boldsymbol{k}_g & \boldsymbol{k}_c \cdot \boldsymbol{k}_g \end{bmatrix}. \qquad (17)$$

Similarly, for $\mathbf{R}_{gh}$. Such matrices are determined from the kinematic BCs. In a general case, they are time dependent and can be obtained also as a function of the Euler angles. What is typically unknown are the positions of O_c and O_h in the fixed frame. In particular their relative position depends on the contact condition and on the presence of friction as explained above. According to Fig. 7, the following vectorial equation can be written

$$O_c O_h(t) = -cl\, \boldsymbol{L_N}(t) / |\boldsymbol{L_N}((t)|, \qquad (18)$$

where $cl = r_c - r_h$ is the radial clearance, very small for hip implants (tens of microns) with respect to both radii. The above equation can be formulated in components in the desired reference frame. In case of frictionless contact, $\boldsymbol{L_N}(t)$ is completely known from the BCs ($\boldsymbol{L} = \boldsymbol{L_N}$), while in case of frictional contact, $L = L_N + L_T$ and an iterative procedure is required to estimate the sliding velocity and identify the frictional force (L_T) direction.

The absolute velocity of a cup/head point is calculated according to the rigid kinematic law

$$v_{P_i} = v_{O_i} + \omega_i \times O_i P_i \quad i = c, h, \tag{19}$$

where ω is the angular velocity of the element. The relative velocity of a cup point P_c with respect to the head, is given by

$$v_{P_c}^r = v_{P_c} - v_{P_h} = v_{O_c} + \omega_c \times O_c P_c - (v_{O_h} + \omega_h \times O_h P_h), \tag{20}$$

where P_c is considered overlapped to P_h. Therefore, the relative velocity of a head point with respect to the cup is

$$v_{P_h}^r = v_{P_h} - v_{P_c} = -v_{P_c}^r. \tag{21}$$

Since the clearance is very small, it can be assumed that $O_h P \approx O_c P$ so that

$$v_{P_c}^r \approx v_{O_c} - v_{O_h} + (\omega_c - \omega_h) \times O_c P_c \approx (\omega_c - \omega_h) \times O_c P_c, \tag{22}$$

where relative velocity of O_h with respect to O_c is considered negligible. It has been checked that for HAs, the above simplification introduces an error <0.5%.

Finally, the sliding velocity v_P^s can be obtained considering the tangential component of the relative velocity, i.e.

$$v_{P_i}^s = v_{P_i}^r - \frac{(v_{P_i}^r \cdot O_i P_i)}{r_i^2} O_i P_i \quad i = c, h. \tag{23}$$

The simplified version of the relative velocity in Eq. (18) can be used to rapidly estimate the sliding velocity.

The sliding velocity is necessary also for determining L_N in frictional conditions, being

$$L_N = L - L_T = L - \sin(\mu)|L|s_v, \quad \text{where} \quad s_v = v_{P_h}^s / |v_{P_h}^s|. \tag{24}$$

Usually, depending on the BCs, the head or the cup centre is considered fixed and the above equation can be further simplified.

4.2.4.3. Wear assessments

The wear is evaluated both as wear depth and wear volume on both bodies. The wear depth h at a cup/head surface point, in a loading cycle of period T, can be obtained adapting Eq. (6) as

$$h(P_c) = \int_0^T \dot{h}(P_c, t) \, dt, \quad h(P_h) = \int_0^T \dot{h}(P_h, t) \, dt, \tag{25}$$

and similarly the wear volumes from Eq. (9)

$$V_c = \int_{A_c} h(P_c) \, dA, \quad V_h = \int_{A_h} h(P_h) \, dA. \tag{26}$$

It should be reminded that, according to Eqs. (3–4)

$$V_c : V_h = k_c : k_h. \tag{27}$$

Since it is also assumed that no significant modification of the contact surfaces occurs (no geometrical up-date), as already mentioned, the wear after n loading cycles can be obtained simply by multiplying h and V produced in a single cycle by n.

4.3. *Wear simulations of MoP HA*

4.3.1. *Effect of wear laws*

Wear investigations presented in [12] were focused on the comparison of the wear laws for UHMWPE discussed in Section 3.4 (Table 1). In particular, the aim was to investigate the role of the cross-shear effect on UHMWPE wear.

Wear maps predicted for a 14 mm implant under *in-vivo* BCs are compared in Fig. 8. All wear laws predict high linear wear in the regions that experience the maximum contact pressure. Qualitatively, the wear maps predicted using the Archard's wear law are very similar independently from k expressions, while a much different wear map is obtained for the new wear law (Eq. 13), which estimates a wider region affected by the maximum wear depth (top of Fig. 8). On the other hand, quantitatively, wear indicators are significantly affected by the wear laws/coefficients with the lowest h_{max} and V values observed for the new wear law (bottom of Figs. 8(a)–8(d)). The results demonstrate the relevance of CS modelling and the necessity to validate wear laws/coefficients by comparing numerical and experimental wear predictions obtained for the same implant, under the same conditions.

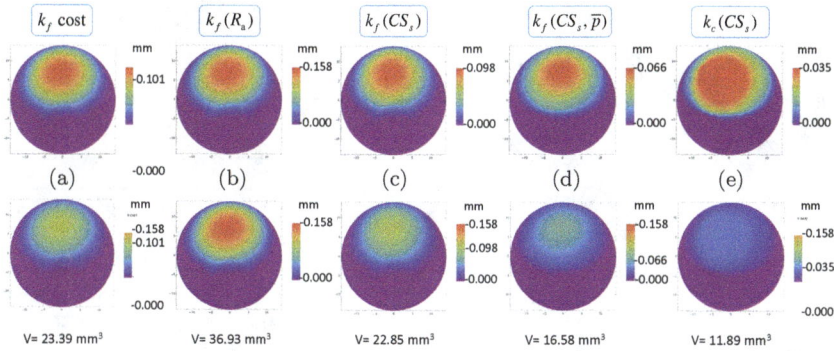

Fig. 8: Linear and volumetric wear at 1 Mc of the plastic cup of a 14 mm MoP HA (in the $x_c y_c$ plane) under *in-vivo* gait conditions, predicted using different wear laws. Maps are drawn both on different scales, within its own minimum and maximum values (top) and on the same scale (bottom). (Reprinted from [12] with permission from Elsevier.)

4.3.2. *Effect of BCs*

The operating conditions of an implant, i.e. loading and kinematics conditions, can greatly influence its wear. This aspect has been investigated in [12]. In particular, a key point was to understand whether the simplified gait conditions assumed in joint simulators for wear testing are suitable to predict the *in-vivo* scenario. Results demonstrate that the implant wear is more sensitive to the kinematic conditions than to the loading ones, and that the wear rates under *in-vitro* BCs underestimate the *in-vivo* ones, with percentage variations of $h_{\max}$ and V up to 44% and 28%, depending on the wear laws (Fig. 9(b)). As an example, Fig. 9(a) and 9(b) show the wear maps obtained for two wear laws in *in-vitro* BCs with results qualitatively and quantitatively different from the correspondent ones obtained under *in-vivo* BCs and portrayed in Figs. 8(d) and 8(e).

Accordingly, results strongly encourage the simulation of more physiological-like BCs in joint simulator wear tests.

4.4. *Wear simulations of MoM HA*

4.4.1. *Effect of friction*

Wear models of MoM HAs have two main peculiarities with respect to the previous ones: bilateral wear and frictional contact.

The presence of friction, as already exposed in Section 4.2.4.2, affects the normal load component and its point of application. In particular, it results in wider trajectories of the nominal contact point K, both on head

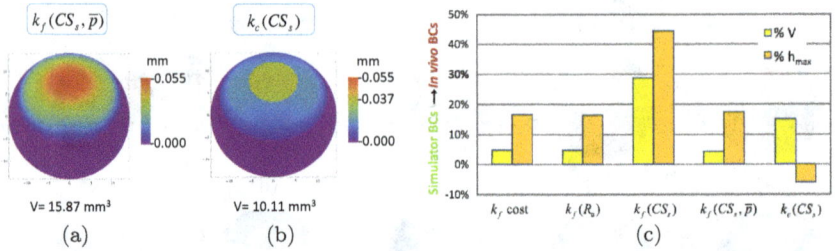

Fig. 9:   Wear maps and volumetric wear at 1 Mc of the plastic cup of a 14 mm MoP HA (in the $x_c y_c$ plane) under joint simulator conditions, predicted using different wear laws (a, b). Percentage variation of h_{max} and V between *in-vitro* and *in-vivo* BCs (c). (Reprinted from [12] with permission from Elsevier.)

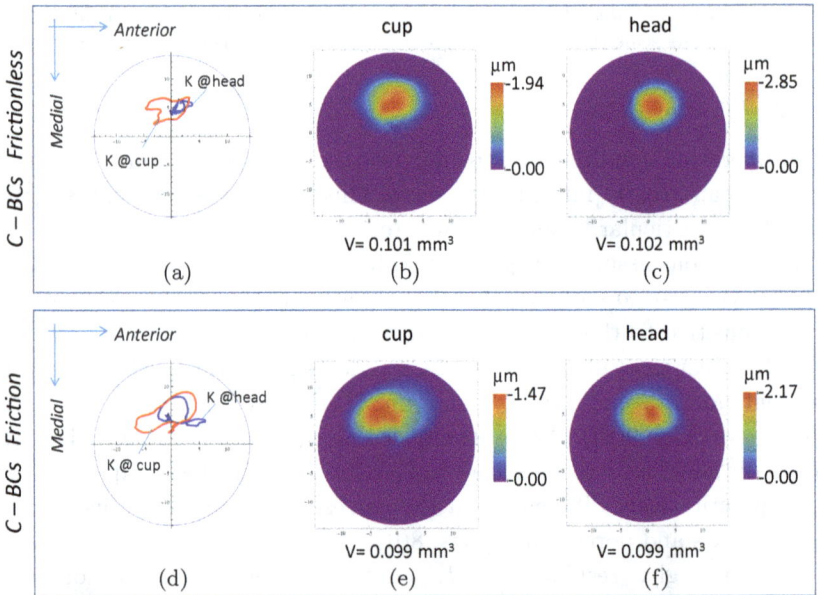

Fig. 10:   Linear and volumetric wear at 1 Mc for a 14 mm MoM HA, under *in-vivo* BCs: frictionless contact (top) vs frictional contact (bottom). (a, d) Trajectories of the nominal contact point K on cup and head; wear map on the cup (b, e) and on the head (c, f). (Reprinted from [24] with permission from ASME.)

and cup surfaces [23, 24]. This is well depicted in Figs. 10(a) and 10(d) that compare the frictionless and friction predictions for a 14 mm HAs under *in-vivo* gait conditions. Accordingly, the friction causes a wider worn area, with lower wear depths, and also a lower volumetric wear (Figs. 10(e), 10(f) *vs.* 10(b), 10(c)).

4.4.2. *Estimation of distinct wear coefficients for head and cup*

Wear models can also be useful for estimating the coefficient k from wear tests directly performed on the components, for example using hip simulators. In this case, Eq. (1) cannot be simply inverted as in Eq. (10), but the general form for the volume loss in Eq. (20) has to be used. Assuming a constant wear coefficient, the linear wear depth becomes

$$h(P_i) = k_i \int_\gamma p(P_i, s) \, ds = k_i \tilde{h}(P_i) \quad i = \text{c}, \text{h}, \tag{28}$$

and the wear volume

$$V_i = \int_{A_i} h(P_i) \, dA = k_i \int_{A_i} \tilde{h}(P_i) \, dA. \tag{29}$$

This expression compared to the experimentally measured volume loss, V^{exp}, provides the formula for k estimation

$$k_i = \frac{V_i^{\text{exp}}}{\int_{A_i} \tilde{h}(P_i) dA} \quad i = \text{c}, \text{h}. \tag{30}$$

This procedure was applied in [9, 25] to estimate the wear coefficients for the head and cup, employing data of distinct wear volumes for the two components available in few literature studies. and used to estimate k_h and k_c, in place of the traditional average of the two frequently used. Results for the case taken from [28] are shown in Fig. 11: k_h is about two-fold k_c and causes the highest linear wear in the head. This is in agreement with the experimental data ($V_\text{h} > V_\text{c}$) and can be explained considering the simulated loading conditions: indeed, as the load was fixed on the head, the head wore more than the cup. Results demonstrate the relevance to use distinct wear factor for the coupling samples in order to obtain reliable wear predictions.

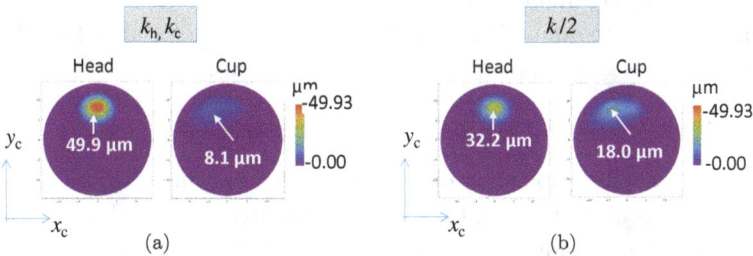

Fig. 11: Linear and volumetric wear at 1 Mc for a 14 mm MoM HA, under *in-vitro* BCs, using specific cup and head wear coefficients (a), or the traditional $k/2$ for both elements (b). (Adapted and reprinted from [9] with permission from Elsevier.)

The novel approach was also applied to evaluate the influence of the implant geometry, i.e. size and clearance, on wear coefficient, both for total and resurfacing implants [9]. By comparing implants subjected to the same BCs, it resulted that the higher the head radius and the lower the clearance, the lower the wear coefficient. Indeed, in both cases the bearing is more conformal and the lubrication is promoted reducing the wear.

Results confirmed the high specificity of the wear coefficient values to the simulated tribological scenario including, in addition to implant materials, implant size and dimensional tolerance, loading and kinematic conditions.

4.5. *Wear simulations of MoP RTSA*

4.5.1. *Estimation of specific wear coefficients for RTSA*

The first objective in wear investigations on RTSAs was to evaluate specific k values for this implant type. Indeed, wear models of RTSAs available in the literature adopt the same values of the wear coefficient for MoP HAs. Given the high specificity of k on the tested conditions, that can cause high error in wear predictions. In [26], experimental wear data on a 42 mm RTSA were used to estimate k using both the Archard and the UHMWPE (Eq. (13)) wear laws. Results demonstrated that the wear coefficients differ significantly for MoP RTSAs and HAs, despite the same coupled materials, and are higher (up to two-fold) for the RTSAs. The computed k were then used to predict wear maps and investigate how they are affected by the wear laws/coefficients. Results are shown in Fig. 12 and are in well agreement with those of MoP HA of Fig. 8: the wear maps predicted by the two wear laws differ significantly, although associated to the same wear volume. The new wear law predicts lower linear wear rates compared to the Archard one, with the maximum wear depth located at the edge of the worn area and not at the cup dome. Consequently, the necessity of a model validation through both wear volume and wear maps is confirmed.

4.5.2. *Effect of size and dimensional tolerance on wear*

The effect of the implant geometry on wear was investigated also for RTSA [27], using the implant-specific k evaluated in [26]. The focus was not only on the implant size (i.e. d_h), but also on the dimensional tolerance. The latter plays a fundamental role in the implant tribological performance

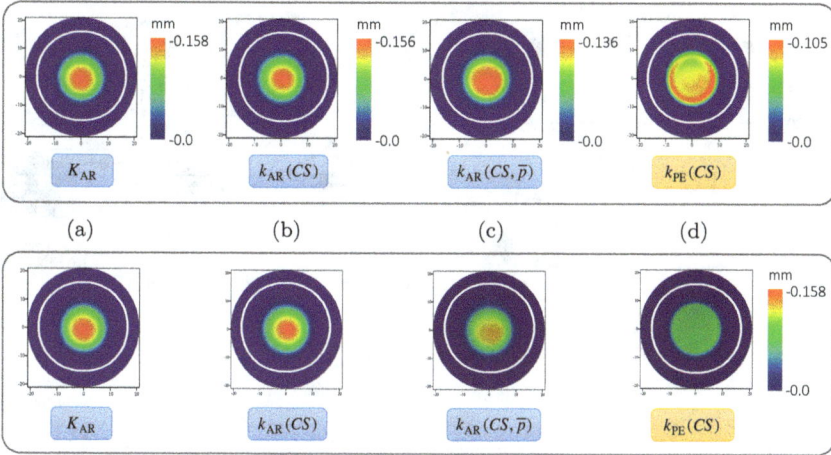

Fig. 12: Linear wear at 2 Mc of the plastic cup of a 42 mm MoP RTSA (in the $x_c y_c$ plane) for a mug-to-mouth task, predicted using different wear laws ($V = 26.6$ mm^3). Maps are drawn both on different scales, within their own minimum and maximum values (top) and on the same scale (bottom). (Reprinted from [26] with permission from Elsevier.)

since it is directly related to the clearance, and thus to the contact conditions and the lubrication regime. However, rather surprisingly, it is not indicated in standards, nor discussed in the literature. A set of 10 different geometries was analysed, considering nominal diameters in the range 36–42 mm, available on the market, and a cup dimensional tolerance of +0.2–0.0 mm [27]. The results of the wear sensitivity analysis to cl and d_h are summarised respectively in Figs. 13 and 14, for different wear laws. They demonstrated that wear indicators are affected by both implant size and dimensional tolerance, although they are more sensitive to the latter. As the implant conformity decreases — for smaller d_h and/or higher cl — the volumetric wear increases, according to both wear laws. On the other hand, the linear wear is only slightly affected by the implant size, while it is highly influenced by the cl. However, different trends are predicted by the two wear laws: according to the Archard's wear law, the higher the cl, the higher the wear depths, in opposite to the UHMWPE law. These investigations remark on the need to validate both wear laws and wear models and the high sensitivity of k to the implant geometry, as well as the necessity to specify stringent limits on the dimensional tolerance in standards.

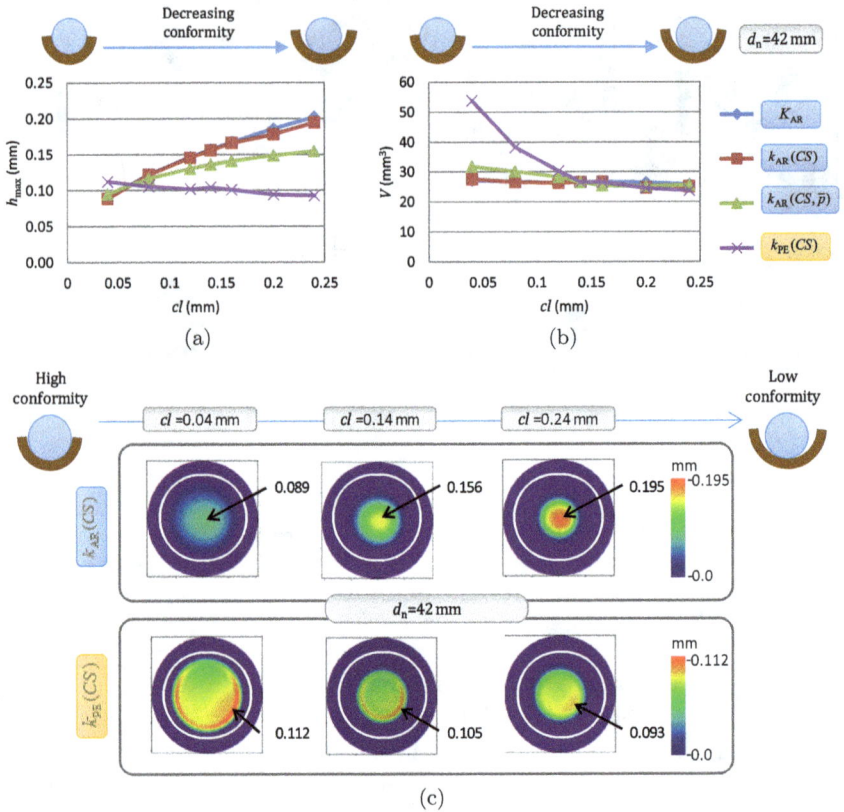

Fig. 13:   Effect of the dimensional tolerance (i.e. *cl*) *cl* on linear wear (a, c) and volumetric wear (b), for a 42 mm RTSA. Comparison of wear predictions for different wear laws and wear coefficients. (Reprinted from [27] with permission from Elsevier.)

4.6. *Wear law/model validity*

The wear law and wear model validity are strictly related and should be assessed by comparing numerical predictions to experimental results obtained simulating the same tribological scenario (e.g. same test samples, operating conditions, etc.). Attention should be paid to this key point. Since the wear coefficient (considered constant) is obtained by comparing V and V^{exp} (Eqs. (10) and (30)), the wear model validation needs the comparison of another wear indicator, i.e. the linear wear over the worn surface, generally named wear map. However, this approach is rarely considered and often a "dependent" model validation is declared based only on wear volumes. One of the reasons is that in some cases wear maps' reconstruction can be

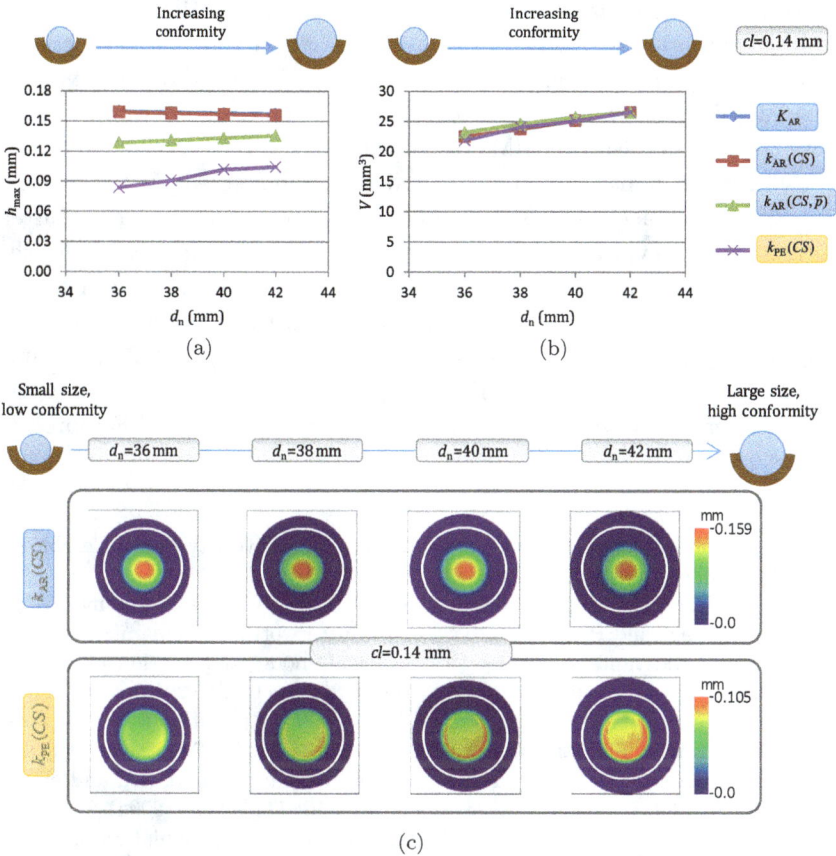

Fig. 14: Effect of the size (i.e. r_{h}) on linear wear (a, c) and volumetric wear (b), for a 42 mm RTSA ($cl = 0.14$ for all implants). Comparison of wear predictions for different wear laws and wear coefficients. (Reprinted from [27] with permission from Elsevier.)

very difficult and costly as it happened for very conformal contacts such as in hip and shoulder implants, for which wear maps are rarely collected. Obviously, the wear law/model validation is even more complicated in case of not constant k (Table 1). Recently, the literature has raised this issue and some attempts at wear maps' reconstruction have been proposed [29].

5. Conclusions

Wear predictive models represent an attractive alternative to expensive and long experimental tests. In recent years, FE simulations have been proposed to investigate wear in mechanical components. However, the analytical approach described in this chapter is still a good option for more rapid indications. The described examples on hip and shoulder implants show the simplicity of the implementation also in presence of complex wear laws, as those needed by UHMWPE surfaces.

The main limit of analytical models is represented by the estimation of contact actions, based on literature formulas (e.g. Hertz, Bartel, Winkler, etc.). Typically, the initial unworn geometry is maintained for the entire simulation, assuming the geometry modifications due to wear can be neglected. This can be true for an initial period, usually correspondent to the running-in. However, volume loss estimated by analytical models are in good agreement also with experimental test results.

Further aspects are discussed in the following Chapter 6 on FE approach.

References

[1] K. L. Johnson, *Contact Mechanics*. Cambridge University Press, Cambridge Cambridgeshire; New York, 1985.

[2] H. R. Hertz, *On Contact Between Elastic Bodies*. Leipzig, Germany, 1882.

[3] I. Hutchings and P. Shipway, *Tribology*. Butterworth-Heinemann, 2017.

[4] J. Lengiewicz and S. Stupkiewicz, Efficient model of evolution of wear in quasi-steady-state sliding contacts, *Wear*. **303**, 611–621 (2013).

[5] L. Mattei and F. Di Puccio, Influence of the wear partition factor on wear evolution modelling of sliding surfaces, *Int J Mech Sci*. **99**, 72–88 (2015).

[6] I. R. McColl, J. Ding, and S. B. Leen, Finite element simulation and experimental validation of fretting wear, *Wear*. **256**, 1114–1127 (2004).

[7] J. Andersson, A. Almqvist, and R. Larsson, Numerical simulation of a wear experiment, *Wear*. **271**, 2947–2952 (2011).

[8] N. H. Kim, D. K. Won, D. Burris, B. Holtkamp, G. R. Gessel, P. Swanson, and W. G. Sawyer, Finite element analysis and experiments of metal/metal wear in oscillatory contacts, *Wear*. **258**, 1787–1793 (2005).

[9] F. Di Puccio and L. Mattei, A novel approach to the estimation and application of the wear coefficient of metal-on-metal hip implants, *Tribol Int*. **83**, 69–76 (2015).

[10] A. Wang, D. C. Sun, S. S. Yau, B. Edwards, M. Sokol, A. Essner, V. K. Polineni, C. Stark, and J. H. Dumbleton, Orientation softening in the deformation and wear of ultra-high molecular weight polyethylene, *Wear* **203–204**, 230–241 (1997).

[11] M. Turell, A. Wang, and A. Bellare, Quantification of the effect of cross-path motion on the wear rate of ultra-high molecular weight polyethylene, *Wear* **255**, 1034–1039 (2003).

[12] L. Mattei, F. Di Puccio, and E. Ciulli, A comparative study on wear laws for soft-on-hard hip implants using a mathematical wear model, *Tribol Int*. **63**, 66–77 (2013).

[13] L. Kang, A. L. Galvin, T. D. Brown, J. Fisher, and Z. M. Jin, Wear simulation of ultra-high molecular weight polyethylene hip implants by incorporating the effects of cross-shear and contact pressure, *Proc Inst Mech Eng H J Eng Med.* **222**, 1049–1064 (2008).

[14] L. Kang, A.L. Galvin, J. Fisher, and Z. Jin, Enhanced computational prediction of polyethylene wear in hip joints by incorporating cross-shear and contact pressure in additional to load and sliding distance: Effect of head diameter, *J Biomech.* **42**, 912–918 (2009).

[15] F. Liu, A. Galvin, Z. Jin, and J. Fisher, A new formulation for the prediction of polyethylene wear in artificial hip joints, *Proc Inst Mech Eng H J Eng Med.* **225**, 16–24 (2011).

[16] T. Maxian, T. Brown, D. Pedersen, and J. Callaghan, 3-dimensional sliding-contact computational simulation of total hip wear, *Clin Orthop Relat Res.* **333**, 41–50 (1996).

[17] T. A. Maxian, T. D. Brown, D. R. Pedersen, and J. J. Callaghan, A sliding-distance-coupled finite element formulation for polyethylene wear in total hip arthroplasty, *J Biomech.* **29**, 687–692 (1996).

[18] M. T. Raimondi, C. Santambrogio, R. Pietrabissa, F. Raffelini, and L. Molfetta, Improved mathematical model of the wear of the cup articular surface in hip joint prostheses and comparison with retrieved components, *Proc Inst Mech Eng H J Eng Med.* **215**, 377–391 (2001).

[19] C. Curreli, F. Di Puccio, L. Mattei, Application of the finite element submodeling technique in a single point contact and wear problem, *Inter J Numer Meth Eng.* **116**, 708–722 (2018).

[20] P. Boileau, D. J. Watkinson, A. M. Hatzidakis, and F. Balg, Grammont reverse prosthesis: design, rationale, and biomechanics, *J Shoulder Elbow Surg.* **14**, S147–S161 (2005).

[21] F. Di Puccio and L. Mattei, Biotribology of artificial hip joints, *World J Orthop.* **6**, 77–94 (2015).

[22] L. Mattei, F. Di Puccio, B. Piccigallo, and E. Ciulli, Lubrication and wear modelling of artificial hip joints: A review, *Tribol Int.* **44**, 532–549 (2011).

[23] L. Mattei, F. Di Puccio, E. Ciulli, Wear simulation of metal on metal hip replacements: an analytical approach, in *11th ASME Biennial Conference on Engineering Systems Design and Analysis, ESDA 2012*, 2012, pp. 555–564.

[24] L. Mattei and F. Di Puccio, Wear simulation of metal-on-metal hip replacements with frictional contact, *J Tribol.* **135**, 111 (2013).

[25] L. Mattei, F. Di Puccio, and E. Ciulli, Estimation of wear factors of metal-on-metal hip implants from simulator tests, in *5th World Tribology Congress, WTC 2013*, Torino, IT, 2013, pp. 413–416.

[26] L. Mattei, F. Di Puccio, T. J. Joyce, and E. Ciulli, Numerical and experimental investigations for the evaluation of the wear coefficient of reverse total shoulder prostheses, *J Mech Behav Biomed Mater.* **55**, 53–66 (2016).

[27] L. Mattei, F. Di Puccio, T. J. Joyce, and E. Ciulli, Effect of size and dimensional tolerance of reverse total shoulder arthroplasty on wear: An *in-silico* study, *J Mech Behav Biomed Mater.* **61**, 455–463 (2016).

[28] S. Williams, D. Jalali-Vahid, C. Brockett, Z. Jin, M. H. Stone, E. Ingham, and J. Fisher, Effect of swing phase load on metal-on-metal hip lubrication, friction and wear, *J Biomech.* **39**, 2274–2281 (2006).

[29] L. Mattei, F. Di Puccio, E. Ciulli, and A. Pauschitz, Experimental investigation on wear map evolution of ceramic-on-UHMWPE hip prosthesis, *Tribol Int.* **143**, 106068 (2020).

[30] ISO 14242-1:2014. Implants for surgery — wear of total hip-joint prostheses Part 1: Loading and displacement parameters for wear-testing machines and corresponding environmental conditions for test.

[31] ISO 14243-1:2009 Implants for surgery Wear of total knee-joint prostheses Part 1: Loading and displacement parameters for wear-testing machines with load control and corresponding environmental conditions for test.

Chapter 6

In-silico Finite Element Wear Predictions

Lorenza Mattei[*,‡], Curreli Cristina[†], and Francesca Di Puccio[*]

*Department of Civil and Industrial Engineering,
Universit di Pisa, Largo Lucio Lazzarino 2, 56126, Pisa, Italy
†Department of Industrial Engineering, Alma Mater
Studiorum — University of Bologna, Via Terracini 24
40131 Bologna, Italy
‡lorenza.mattei@unipi.it

The recent remarkable improvements in software and computer technologies have enabled the development of complex Finite Element (FE) wear models, that find application in many different fields from automotive to biomedical engineering. The strength of FE wear models is their capability to predict how wear, modifying the geometry and the contact actions at the interfaces, affects the fitness for use of a component during its lifetime. Additionally, since it is not easy to replicate experimentally the working condition of a component, numerical simulations represent a valuable support in the design phase.

The chapter presents a general overview on FE wear modelling procedure, that is typically based on three main steps: contact analysis, wear calculation according to a specific wear law, wear implementation plus an eventual mesh update, repeated sequentially for each instant of the loading history. The wear calculation and implementation can be performed through complex and versatile user-defined subroutines or, more easily, by using automatic software tools.

One of the main drawbacks of this approach is the high computational cost, that however can be reduced by means of specific strategies and modelling methods. An accelerated procedure for cyclic loading and the submodelling technique for non-conformal contacts are proposed and their application to pin-plate/pin-disc wear models is described in detail. Their capability to reduce computational costs without compromising the accuracy is demonstrated.

1. Introduction

Finite Element (FE) models can be considered a gold standard in mechanical design as well as in many other fields. In the last decades, the complexity of models and simulations has rapidly increased, taking advantage of improvements in software and technologies.

Wear modelling involves many aspects also at different scales, from micro to macrogeometries. A general overview of the problem is described in Chapter 5, where a simple analytical approach is presented, which is useful for a first approximation of the solution when the contact problem can be solved analytically and the geometry modification due to wear is negligible. In a more general case, in particular in non-conformal contacts, it is important to estimate the evolution of the contact pressure and surface geometry as wear progresses. That can be conveniently investigated by FE models, using predefined tools or user-subroutines.

In this chapter, the main points of FE wear simulations are discussed with some examples, first examining steady-state boundary conditions (BCs) and secondly, cyclic loadings. The major limit of these analyses is their computational cost, which can be reduced by the application of the submodelling technique, introduced in the last part of the chapter.

2. FE Wear Models: General Aspects

When adopting a Finite Element approach, the main steps of wear simulations include (Fig. 1):

(a) contact analysis, to determine the contact pressure and the sliding distance between the surfaces in contact;

(b) wear calculation, for each node the increment of wear depth is calculated according to the adopted wear law, most frequently the Archard's wear law. This step can be performed through an automatic software tool or through a user-defined routine, e.g. the UMESHMOTION in Abaqus®;

(c) wear implementation and mesh smoothing; the calculated wear depth is applied to the mesh as a displacement of the nodes, so that geometry is modified. In order to avoid excessive elements distortion or collapse, a check on their shape is performed and, if needed, the mesh is improved with an adaptive mesh smoothing.

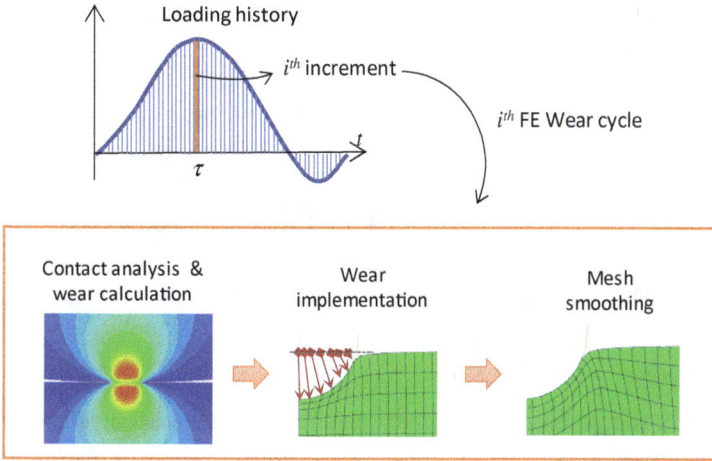

Fig. 1: Basic steps in wear simulations of generic loading history. (Adapted and reprinted from [4] with permission from Elsevier.)

These steps are repeated iteratively to simulate a loading history, i.e. a specific operational time of the component. Thus, a complete simulation can require hours or days. As already stressed, the computational cost remains the main drawback of wear FE analyses [1–4], therefore analysis settings should be accurately chosen to achieve reliable results in a reasonable time.

2.1. Wear calculation

As mentioned above, wear calculation can be performed through a specific tool of the FE software or by user-defined routines called by the software itself. That depends both on the software and the adopted wear law (see Chapter 5, Section 3). For example, in Ansys® APDL and Workbench the generalised Archard's wear law is predefined; accordingly, the wear depth increment Δh occurring during the time interval Δt is evaluated as

$$\Delta h = \frac{K}{H} p^m \left| v_s \right|^n \Delta t, \tag{1}$$

where K is the wear factor, H the material hardness, p the contact pressure, v_s the sliding velocity so that $\left| v_s \right| \Delta t$ represents the sliding distance increment.

On the other hand, the implementation of more complex wear laws, as those based on the cross-shear described in Chapter 5, typically requires user-defined subroutines, with properly defined input (e.g. stress and strain fields, sliding distance increment components) and wear increment and

direction as output. This option is available, for example, both in Ansys®
and in Abaqus®.

In user subroutines, the discrete form of the wear increment at the node
j at a given iteration i is expressed using the average pressure between two
consecutive time steps and the sliding increment Δs, i.e.

$$\Delta h_i^j = \frac{K}{H} \left(\frac{p_i^j + p_{i-1}^j}{2} \right)^m \left(\frac{s_i^j - s_{i-1}^j}{\Delta t} \right)^n \Delta t. \tag{2}$$

Sometimes the average pressure is replaced by the actual pressure value.

2.1.1. *Implicit or explicit kinematics*

In general, the sliding distance/velocity at each node is calculated during
the simulation, specifying the relative motion of the parts in contact as a
BC. This is sometimes referred to as explicit kinematic approach. As an
example, consider the pin-on-plate wear test shown in Fig. 2. The pin can
only translate towards the plate under the action of the load $\boldsymbol{F}$, while the
plate translates horizontally, specifying its position in the fixed frame at
each time step. That implies that the meshes of the contact surfaces have
nodes differently misplaced at each time step. This can produce some minor
"irregularities" in contact pressure profile.

In some particular cases the alternative implicit kinematics formulation
can be adopted, i.e. the simulation of the relative motion of the parts can
conveniently be omitted, sparing computational time. For instance, it can
be done when the relative motion is translatory, the contact is frictionless
and only one body (the pin in the example) gets worn.

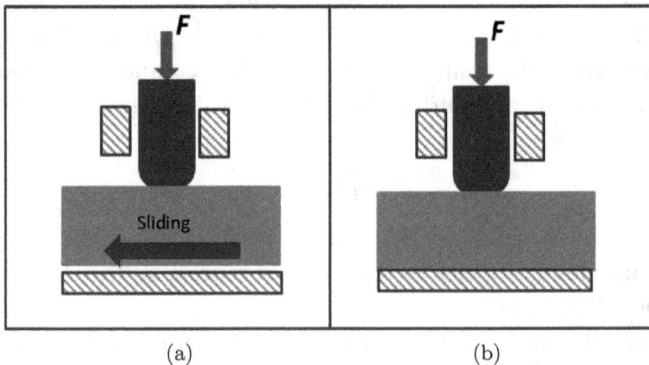

(a)   (b)

Fig. 2:   Pin-on-plate scheme with explicit (a) or implicit (b) kinematic approach.

This consists in combining a static analysis and a modified wear law with a fictitious wear coefficient k^*, i.e.

$$\Delta h = k^* p^m \Delta t \quad \text{with} \quad k^* = \frac{K}{H} |v_s|^n. \tag{3}$$

For the case of the pin-on-plate, simulations can be performed also fixing the plate in a static configuration and calculating wear through the modified law of Eq. (3).

2.1.2. *Unilateral wear or bilateral wear*

In wear simulations, an important distinction is usually made between two cases:

(a) unilateral or asymmetric wear, when only one body undergoes wear, and
(b) bilateral or symmetric wear, when both bodies get worn (Chapter 5, Section 3.1). The distinction depends on the wear resistance of the rubbing surfaces. In many cases, e.g. brakes, it is a design choice to have only one element subjected to wear. In other cases, e.g. gears, wear is bilateral.

In numerical simulations, bilateral wear is somehow more complex, since it is often solved as a combination of two unilateral cases. That also implies an increase of the computational time.

Another point should be mentioned about bilateral wear: a wear law is required for each surface of the coupling. In case of the Archard's wear law, it means that a wear coefficient $k = K/H$ for each material must be inserted as input. However, since wear is a phenomenon characterising the whole coupling separate indications for the coupling elements are rare [1, 2, 4]. Often the total volume loss is measured and a simple equi-repartition is supposed, meaning the same K/H value is adopted for both materials [1].

2.2. *Mesh definition*

When defining the mesh of a wear model, the same indications used for the contact analysis hold. The element size should be properly selected considering the characteristic dimension of the contact area, that means having very small elements for non-conformal contacts. However, two points are worth remembering, as further detailed in Sections 3 and 4:

(a) Wear increases the conformity of the contact surfaces, causing an increment of the contact area, with a consequent reduction of contact pressure. Thus, a coarse mesh that initially approximates rather roughly the pressure distribution could be good to catch its evolution with wear.

(b) Mesh smoothing settings are usually based on the reduction of the element volume with time, as wear proceeds. For example, assuming that the threshold is a 50% volume reduction, small elements require mesh smoothing more often than larger elements, that means also longer computational times. Thus, a good compromise is fundamental when choosing the elements size in wear models.

2.3. *Model validation*

Model validation is a crucial point in all the applications. A comparison with reliable experimental results should be performed to assess the accuracy of the simulations. As already stressed in Chapter 5, wear models are usually validated considering only the wear volume, but the wear map is fundamental as well. The reason is double: on the one hand, the volume loss is also used to evaluate K/H, on the other hand, the same wear volume can be produced by different wear maps. Consequently, only wear map validation would confirm the reliability of both the model and the wear law adopted to describe the phenomenon.

However, since wear is a very complex phenomenon, dependent on many factors, wear test results usually have a wide dispersion. Therefore, one should not aim at a "perfect" correlation of numerical and experimental predictions but should be satisfied with a good agreement in the range of the experimental dispersion.

3. Simulation of Cyclic Conditions

3.1. *General procedure*

In many applications, wear is caused by periodic loading conditions, such as in pin-on-plate sliding tests or gait cycle wear tests for hip prostheses. Consequently, experimental and numerical wear investigations typically need to simulate a very high number of wear cycles, e.g. up to 5 Mc (million cycles) for wear in hip prosthesis.

FE wear simulations of cyclic conditions performing a quasi-continuous geometry update could have a very high computational cost. As a schematic

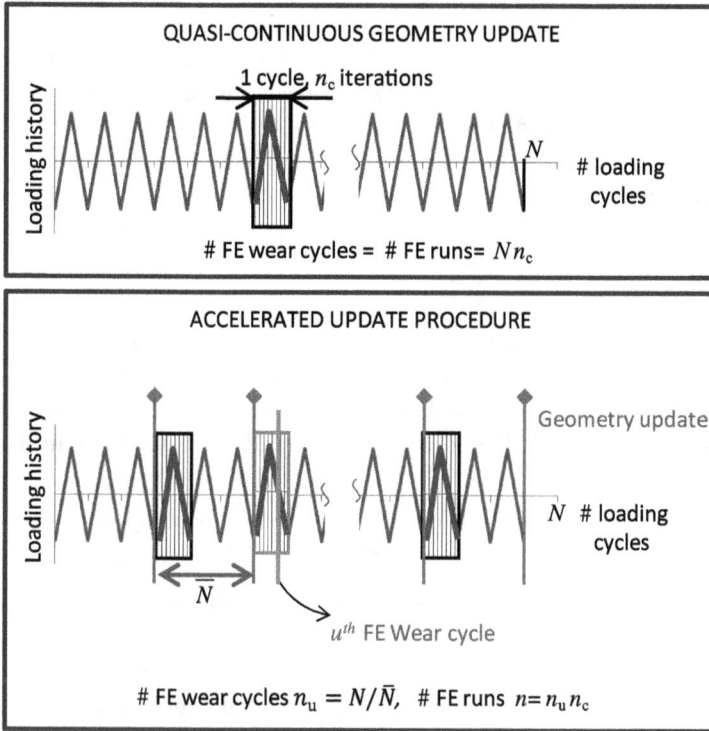

Fig. 3: Procedures for wear simulations under cyclic conditions: quasi-continuous vs. accelerated geometry update procedure. (Adapted and reprinted from [4] with permission from Elsevier.)

example, consider a periodic loading history characterised by N loading cycles, shown at the top of Fig. 3. Assuming each wear cycle discretised in n_c increments, the simulation of the whole loading history would require a number $n = Nn_c$ of contact analyses (i.e. FE runs) and geometry updates (i.e. FE wear cycles). This means that, for instance, in case of FE wear cycles discretised in 25 increments each one processed in 30 s, the simulation of 1000 cycles would last about nine days. That points out the key role of simplifying hypotheses and computational strategies in reducing the duration of wear simulations. For this purpose, the accelerated update procedure illustrated at the bottom of Fig. 3 has been demonstrated to be very effective [2, 4–8]. Basically, it hypothesises that wear affects the contact pressure only after a critical number $\bar{N}$ of cycles because the wear depth in a wear cycle is usually very low. Consequently, the geometry update can be performed every $\bar{N}$ cycles, allowing to extremely accelerate the wear simulation that needs a lower number $n_u = N/\bar{N}$ of FE wear cycles, and a lower number $n = n_u n_c$ of FE runs. In fact, the parameter $\bar{N}$ is also known as accelerated factor [5]. Accordingly, for each wear cycle u (Fig. 3), the local wear depth at the node j accumulated during a given wear cycle u is calculated by applying the accelerated factor $\bar{N}$ as it follows, assuming the Archard's wear law:

$$\Delta h_u^\kappa = j\,\overline{N} \sum_{i=1}^{n_C} \frac{\left(p_{u,i}^\kappa + p_{u,i-1}^\kappa\right)}{2} \left(s_{u,i}^\kappa - s_{u,i-1}^\kappa\right), \tag{4}$$

where i is the generic increment within the wear cycle. It is worth noting that although $\bar{N}$ is typically considered constant during wear evolution, it strongly depends on the simulated tribological conditions. Moreover, the correct setting of both $\bar{N}$ and n_c needs a critical sensitivity analysis [4].

3.2. *Application to cylinder-on-plate reciprocating wear tests*

This section provides an example of the application of the accelerated procedure presented by the authors in [4]. The case study consists of a cylinder-on-plate under reciprocating contact, with the rubbing surfaces having different wear resistances, as shown in Fig. 4(a). This relatively simple model was exploited to investigate:

(a) Model converge criticality for the accelerated procedure;
(b) The evolution of the contact and wear parameters during wear progress;

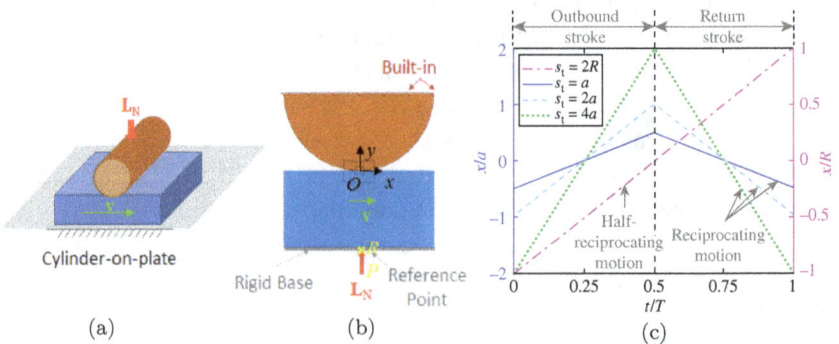

Fig. 4: Modelling of cylinder-on-plate wear tests (a) and (b) under cyclic conditions characterised by periodic reciprocating sliding contact under a constant load (stroke amplitude, s_t; Hertzian semi-width a) (c). (Adapted and reprinted from [4] with permission from Elsevier.)

(c) The effect of the partition factor α (Eq. (5)) in Chapter 5), i.e. wear redistribution, on wear parameters;

(d) The effect of the stroke amplitude s_t, on wear parameters.

3.2.1. *Model description*

A 2D (plane strain) wear model was developed in Abaqus® using the UMESHMOTION subroutine, *ad hoc* coded in FORTRAN.

The model consists of a cylindrical pin of radius $R = 10\,\text{mm}$ coupled with a plate of length $2R$, both made of metal. The non-conformal contact is assumed to be frictionless and is set as double, allowing the simulation of bilateral wear. The cyclic loading conditions consist in reciprocating sliding contact under a constant load of $100\,\text{N}$. As shown in Fig. 4(b), the top surface of the pin is built in, while the kinematics of the plate is driven by the reference point (RP) of a rigid base tied to the bottom of the plate so that rotation is constrained, y translation is controlled by the load applied to the rigid base, and x translation is used to reproduce the reciprocating contact, as described in Fig. 4(c). Consequently, an explicit kinematics formulation is adopted. In particular, both fretting (reciprocating motion) and sliding wear (half-reciprocating motion) are simulated by varying the stroke amplitude s_t, for a total of four conditions: $s_t = a, s_t = 2a, s_t = 4a, s_t = R$, where $a = 0.11\,\text{mm}$ is the hertzian semi-width in unworn status.

Although contact bodies are both metallic, different wear resistances of the rubber surfaces are simulated assuming α values of 0 (no wear of pin), 0.25, 0.5, 0.75, 1 (no wear of plate), and the same couple wear coefficient $k = 10^{-8}\,\text{mm}^2/\text{N}$.

3.2.2. *Results*

3.2.2.1. Model convergence analysis

The convergence analysis of wear models assuming the accelerated procedure is very critical. It has a double face, as it must be performed both for the unworn configuration, i.e. classical mesh sensitivity analysis, and the worn configuration, i.e. setting of $\bar{N}$ and n_c. Note that a too high $\bar{N}$ or a too low n_c could cause numerical instabilities such as local peaks of contact pressure and unsmooth worn surfaces, while the contrary could slow down the simulations. Consequently, once the mesh for the unworn condition is defined, the sensitivity analysis on $\bar{N}$ and n_c allows to identify values that guarantee a good compromise between accurate results and computational cost.

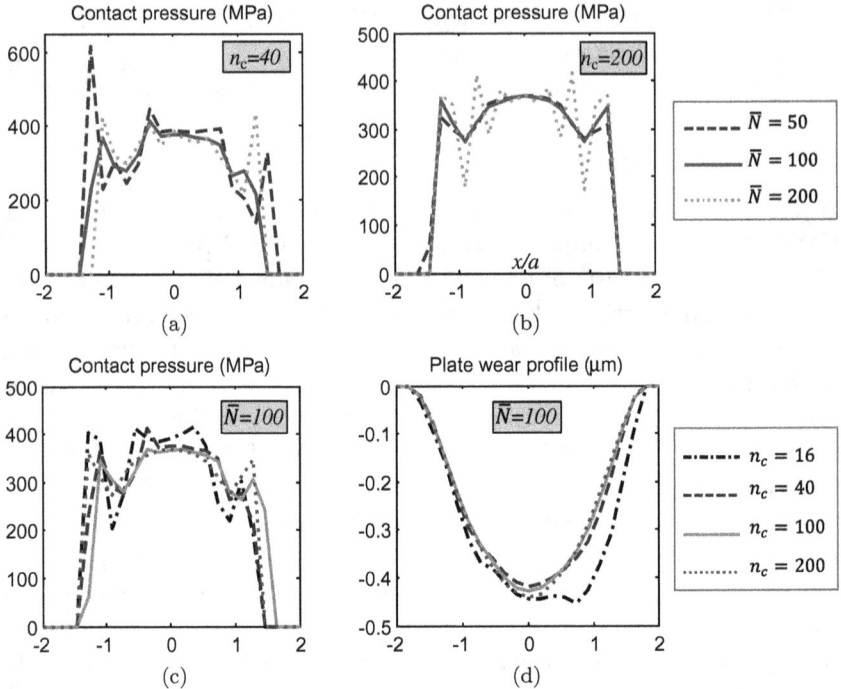

Fig. 5:   Model convergence analysis for simulations of cyclic conditions: effect of $\bar{N}$ and n_c on numerical instabilities in the prediction of the contact pressure (a)–(c) and the worn profiles (d) obtained for $s_t = a$ and $\alpha = 0.5$, at $N = 1000$. (Adapted and reprinted from [4] with permission from Elsevier.)

Figure 5 shows some results of the convergence analysis for the cylinder on plate model, for the case $s_t = a$ and $\alpha = 0.5y$ [4]: the contact pressure distribution (Fig. 5(a)–5(c)) and the plate wear profiles (Fig. 5(d)) are compared for different combinations of $\bar{N}$ and n_c values. Low values of n_c (<100) cause local contact peaks, independent of $\bar{N}$ values, while, high values of n_c ($\geq$100) can produce smooth and symmetric contact pressure profiles when combined with quite high $\bar{N}$ values ($\bar{N} \geq 100$). Note that lateral peaks are expected as caused by discontinuities in curvature radius at the contact edges. Low values of n_c also cause unsmooth and asymmetric wear profiles, as shown for the plate in Fig. 5(d). According to the convergence analysis, the combination $n_c = 100$ and $\bar{N} = 100$ guarantees both smooth contact pressure and wear profiles, as shown at the bottom of Fig. 5.

3.2.2.2. Evolution of contact and wear parameters

The typical results of an FE wear model consist in the evolution of contact pressure and wear profiles during the loading history. An example is provided in Fig. 6 for the case $s_t = a$ and $\alpha = 0.5$. In unworn conditions (i.e. at the first load cycle) the contact is described by the Hertzian theory. Later, as the surfaces wear out, the contact becomes more conformal: a wear scar develops on the pin surface while a wear groove hollows the plate surface out, the contact area widens, and the contact pressure flattens. The contact pressure at the centre of the worn area, referred to as maximum contact pressure, decreased rapidly in the first 1000 wear cycles. This trend is reflected in the evolution of the maximum wear depth of both pin and plate surfaces, which increase faster in the first phase of the wear process.

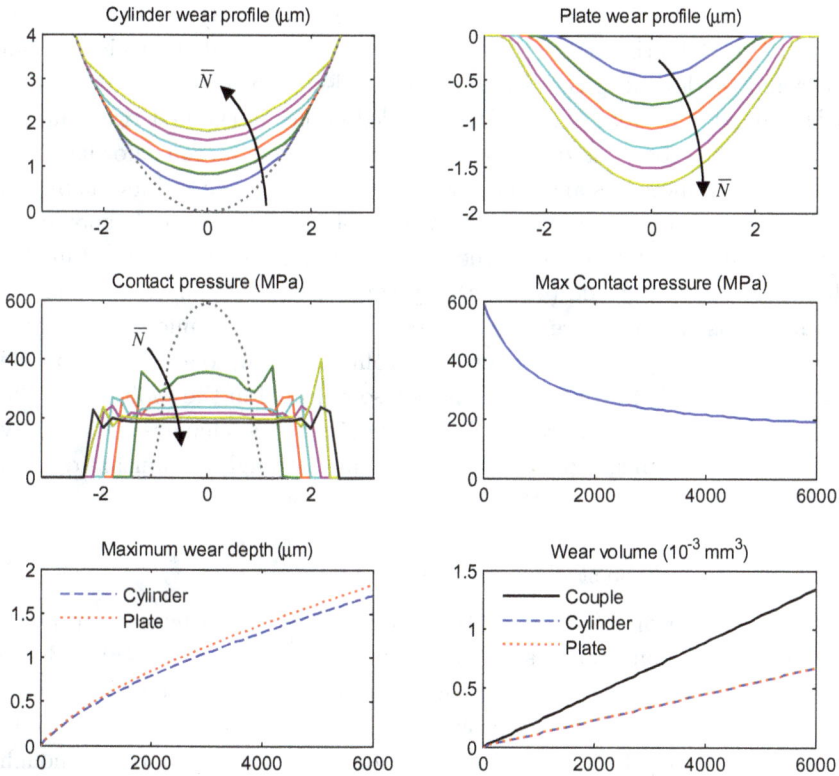

Fig. 6: Evolution of contact and wear parameters for the case $s_t = a$ and $\alpha = 0.5$, over 6,000 wear cycles. (Adapted and reprinted from [4] with permission from Elsevier.)

On the contrary, in agreement with the theory, the trends of wear volumes (single body and couple) are linear and in particular, being $\alpha = 0.5$, equal wear volumes are predicted for pin and plate.

3.2.2.3. Effect of the wear partition factor

As largely discussed in Chapter 5, the different wear resistance of the rubbing surfaces, though rarely discussed and modelled in literature, is a key aspect of the wear process.

The pin-cylinder model is used to demonstrate the important effect of the partition factor on wear evolution, by assuming α values in the range 0–1. Both unilateral ($\alpha = 0, 1$) and bilateral ($\alpha = 0.25, 0.5, 0.75$) wear conditions are considered and results obtained for two stroke amplitudes are reported in Fig. 7.

Moving from the case of unilateral wear of pin ($\alpha = 0$) to unilateral wear of plate ($\alpha = 1$), the higher the α value, the more marked and wider the pin scar and the less deep the plate groove, as also confirmed by the evolution of the maximum wear depth of the coupled bodies in the first two columns in Fig. 7. Moreover, as α increases, the contact gets more conformal: The contact area increases and the contact pressure flattens, and its maximum value drops down rapidly during wear evolution. However, it should be noted that the effect of α on the contact pressure is more important for higher stroke amplitude ($4a \, vs. \, a$). By varying the value of α, the contact pressure distributions are very similar for $s_t = a$ while much different for $s_t = 4a$, moving from a quasi-hertzian profile at $\alpha = 0$, to a flattened profile for $\alpha = 1$, Fig. 7 third column. This is also reflected in the evolution of the maximum contact pressure which, in case of $s_t = 4a$, changes significantly passing from a linear ($\alpha = 0$) to a faster and strongly nonlinear ($\alpha = 1$) decrease.

3.2.2.4. Effect of the stroke amplitude

The simulation of different stroke amplitudes allows to investigate much different wear conditions, such as the fretting ($s_t = a, 2a, 4a$) and the sliding wear ($s_t = 2R$). Also in this case, results are presented in terms of wear and contact pressure profiles, for different values of α, as shown in Fig. 8. The effect of the kinematic conditions is certainly important, though variable with α. In case of bilateral wear or unilateral wear of pin (first two rows), the effect of s_t is comparable to that of α: the higher the s_t, the less deep the plate groove and thus less conformal the contact, with higher contact pressures (less flattened). However, it should be noted that the

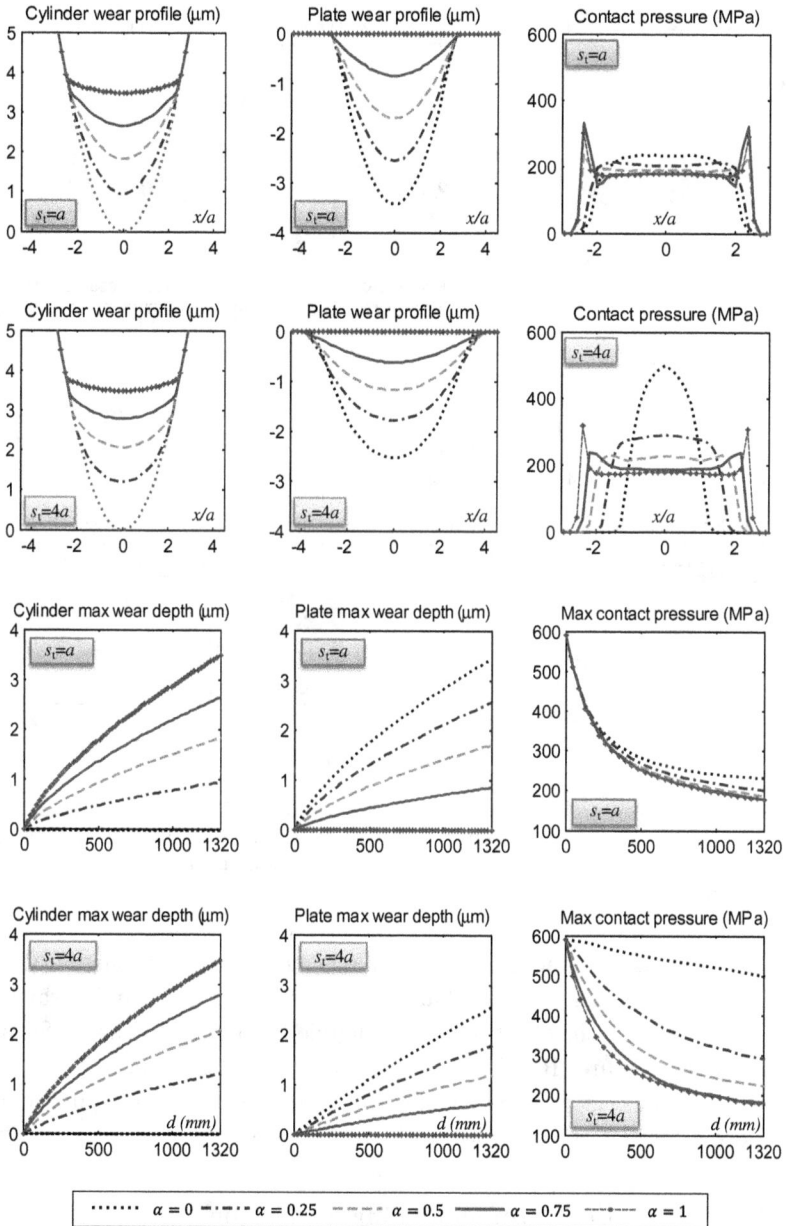

Fig. 7: Effect of the wear partition factor α on cylinder and plate wear profiles and contact pressure (for a sliding distance of 1,320 mm) and the evolution of maximum wear depth and contact pressure. (Adapted and reprinted from [4] with permission from Elsevier.)

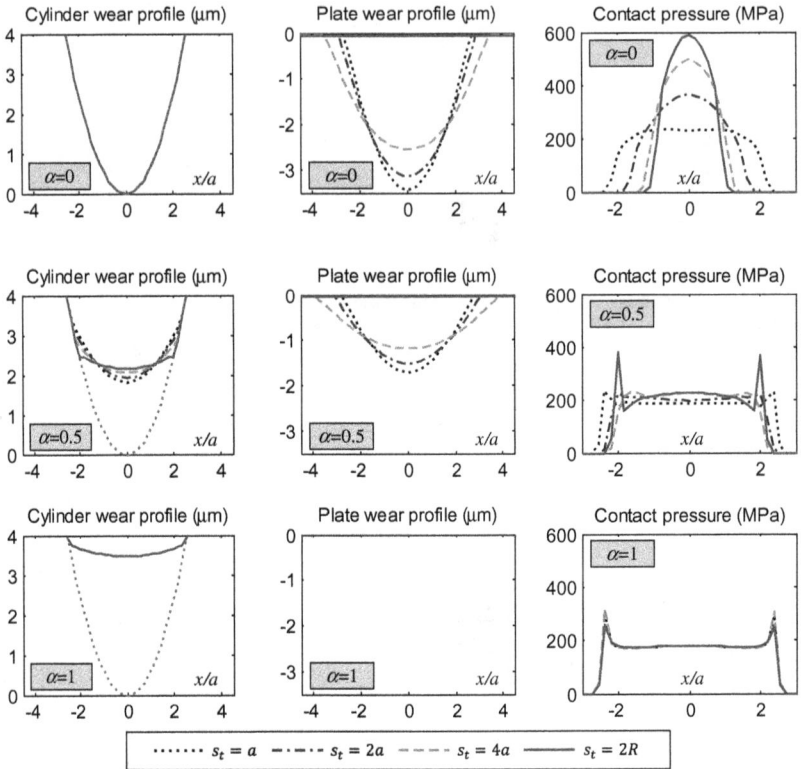

Fig. 8: Effect of the stroke amplitude s_t on wear profiles (cylinder in left column, pin in central column) and contact pressure (right column) for a sliding distance of 1,320 mm. Specific values of the wear partition factor α are considered in each row. (Adapted and reprinted from [4] with permission from Elsevier.)

effect of s_t on the contact pressure profile is more marked for higher values of α (i.e. 0.5 *vs.* 0). On the other hand, in the extreme case of unilateral wear of the pin ($\alpha = 1$), the effect of s_t is negligible both on the wear profile and the contact pressure. Results demonstrate that fretting and sliding wear cause much different wear damage, with maximum contact pressure higher and potentially more dangerous in the latter case.

4. Application of the Submodelling Technique

4.1. *General aspects and procedure description*

The submodelling technique is one of the methods that can be used to reduce the computational cost problem of FE wear simulations. Compared

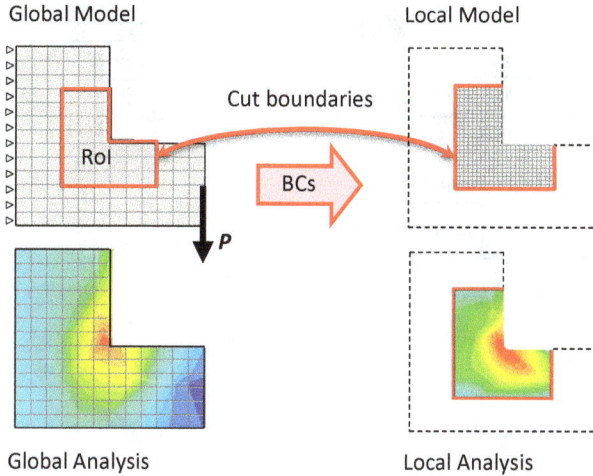

Fig. 9: Submodelling in FE analysis. The solution of the coarse global model is used to provide the BCs to a fine local model that includes the RoI delimited by the cut boundaries. The accuracy of the results is improved in the local model. (Reprinted from [3] with permission from Elsevier.)

to other strategies proposed to minimise the calculation time, mainly based on extrapolation methods [9, 10], this technique proved to be an efficient solution also when few loading cycles are considered [3].

The main idea behind the submodelling approach for generic FE models is that the solution of a coarse-mesh-model of the entire system, usually called *global* model, can be used to provide the BCs to a fine-mesh-model (submodel or *local* model) that only includes the region of interest (RoI). Nodal displacements, forces or stresses are usually extracted from the global model solution at the cut boundaries and then applied to the local model (Fig. 9). The solution obtained using the combination of global and local models is accurate and usually faster than the one achieved with the entire detailed model.

When applying the submodelling technique to wear models, the general workflow depicted in Fig. 10 can be adopted [3]. Three main important steps can be identified: development of the global model, development of the local model and wear simulations.

(a) *Global model development*

The global model is solved to provide the BCs to the local model used to simulate the wear test. Two important aspects of the global model development should be carefully evaluated. The first one regards the

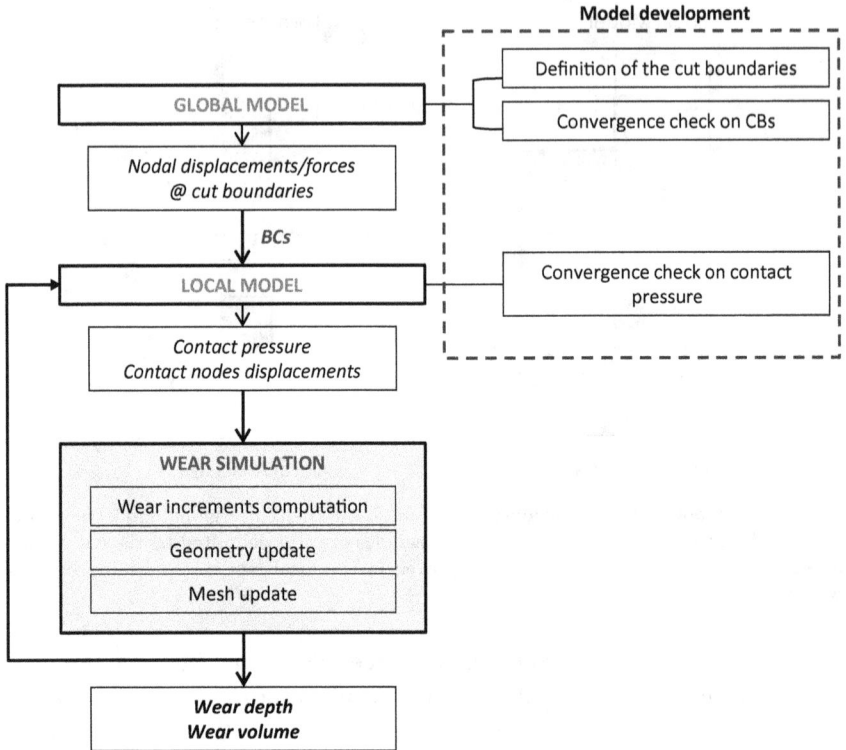

Fig. 10:   Main steps of the FE wear submodelling procedure: (a) global model development, (b) local model development and (c) wear simulations. (Reprinted from [3] with permission from Elsevier.)

definition of the cut boundaries. Submodelling is based on St. Venant's principle and general recommendations suggest placing the cut boundaries far enough away from the stress concentration region. In case of wear analyses, this region varies during the simulation because the contact area increases due to wear. The cut boundaries should be thus defined so that they can be considered far enough away from the stress concentration region for the entire wear simulation. The second important aspect for global model development is related to the "acceptable" mesh density. Despite the fact that contact pressure is usually not an output of this model and can be roughly estimated using a coarse mesh, a convergence analysis on the quantities to be transferred (i.e. nodal forces, displacements or stresses at the interfaces) is needed in order to reduce the so-called "BCs error" [11]. The element size at the contact region of the global model should be thus properly selected so that the "right" behaviour at the interfaces of the system is captured.

(b) *Local model development*

The first critical step of this model development phase is the definition of the BCs. The most common submodelling approach in structural mechanics is based on the use of nodal displacements. However, this strategy cannot be purely applied for wear analyses because the geometry significantly changes after the material removal. Alternative stress and force-based techniques can be evaluated, eventually combined with the displacement-based one in a hybrid scheme. Choosing the right mesh density to accurately predict contact stresses and wear is a second important aspect in the local model development. A mesh convergence analysis aimed at reducing the discretisation error on the maximum contact pressure is proposed in the wear submodelling procedure. However, general considerations on the effect of the element size on the wear simulation results (Section 2.2) should be taken into account in this phase.

(c) *Wear simulations*

The main steps of wear simulations (Section 2) are usually performed entirely with the local model while the global model is used to simulate a simple contact analysis in the initial unworn configuration. This is the case of the single point contact example that will be presented in the following section and used to discuss the validity of the wear submodelling procedure. However, it is worth noting that in case of more complex wear problems (e.g. multipoint contact and wear analyses where the total contact force acting on the contact surfaces might change during the first simulation phase as a result of different contact geometries), the global model can be used to simulate the first wear cycles in order to correctly estimate the BCs at the model interfaces.

The general wear submodelling procedure is here described in its simplest formulation. In fact, the global model is used only for the initial unworn configuration and the BCs are transferred to the local model and kept constant for the entire wear test. Remember the important assumption that the BCs at the selected interfaces do not vary significantly with wear/time since the cut boundaries are set far enough away from the stress concentration region for the entire wear simulation. However, in general, the consistency of this assumption should be always checked. To this aim, contact analyses performed on the global model with reasonably expected wear profile might help in defining the cut boundaries and assess model assumption. On the other hand, for the cases where BCs at the cut boundaries evolve with wear, new global analyses with the updated worn geometry are needed, thus performing global-local model iterations. This can occur, for example, in a multi-point contact model, where the force distribution in the contact areas varies with wear as the variation of geometry causes a modification of the contact stiffness, as described in [12].

4.2. Validation of the submodelling procedure

In this section, the validity of the wear submodelling procedure is discussed for the case of a single point contact. In particular, the case study presented in [3] is considered, which consists of a pin-on-disc wear test where a metal hemispherical head pin is in contact with a metal rotating disc under a constant normal load (Fig. 11).

4.2.1. Models' description

Two-dimensional axisymmetric models were developed in Ansys® mechanical APDL. The global model consisted in a quarter of circle of radius R and a quadrilateral region of dimension $R \times R$, with $R = 5$ mm. A constant load of 21 N is applied at a reference node at the bottom rigid surface of the disc while the top surface of the pin is built in (Fig. 11).

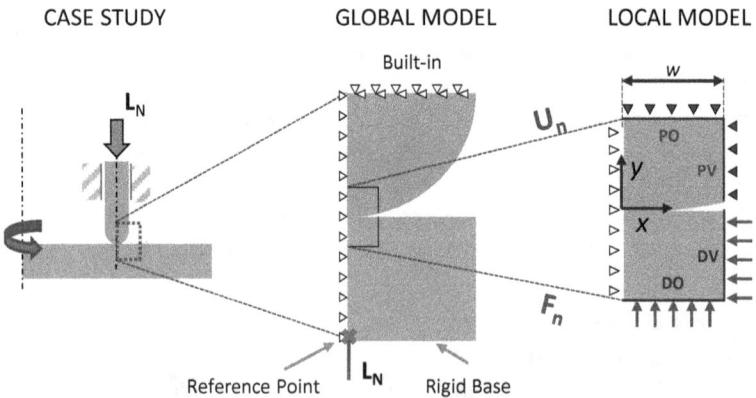

Fig. 11:   Case study on f pin-on-disc. BCs and geometries of the global and local model. Nodal forces F_n and nodal displacements U_n were extracted from the global model solution and mapped, respectively, to the disc and pin boundaries (DO, DV, PO and PV). (Adapted and reprinted from [3] with permission from Elsevier.)

The local model consists of the subregions introduced in the global model and delimited by the cut boundaries of length $w = 1\,\mathrm{mm}$ (about 10 times the Hertzian contact half-width, a, in the unworn condition). A hybrid submodelling approach is adopted where nodal forces and displacements extracted from the global model solution are mapped, respectively, to the disc and pin boundaries. The wear test is simulated using the local model for a total sliding distance of $3\,\mathrm{m}$.

For both global and local models, the contact is simplified as frictionless and the material loss is attributed only to the pin. The kinematics is implicitly modelled by introducing a fictitious wear coefficient $k^* = kv$, where k is the wear coefficient set to $1.25 \times 10^{-13}\,\mathrm{Pa}^{-1}$ and v is the sliding velocity, assumed uniform in the contact nodes and equal to $25\,\mathrm{mm/s}$. The wear model was implemented in the software using the TBDATA, WEAR command with the Archard's wear law option.

4.2.2. *Results*

4.2.2.1. Convergence analysis of the global model

The global model convergence analysis is a fundamental step of the procedure. As already stressed, the boundary error on the quantities to be transferred to the local model must be minimised by properly selecting the mesh density. Figure 12 shows some results on the influence of the edge element size (h_G) near the contact on the nodal displacement profiles at the DO and DV boundaries defined in Fig. 11. Slightly different solutions were obtained varying the element size with discrepancies that were reduced with the mesh refinement. The "right enough" mesh size should be obviously defined based on the acceptable convergence criteria for the desired solution accuracy. In this case, a coarse mesh with $h_G = 1/3a$ was selected

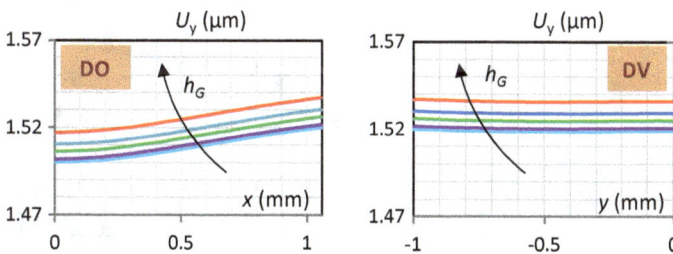

Fig. 12: Variation of the displacements U_y profiles along the cut boundaries DO and DV with a decreasing value of the element edge size (h_G) near the contact region of the global model. (Adapted and reprinted from [3] with permission from Elsevier.)

because percentage differences on both displacement and nodal stresses at the cut boundaries were always lower than 1%.

4.2.2.2. Definition of the cut boundaries

The size of the subregions was selected based on two opposite requirements: a small value of w to reduce the computational cost and a large value of w to guarantee that, at the end of test, the cut boundaries were still far enough away from the stress concentration region. A value of $w = 1\,\mathrm{mm}$ proved to be a good compromise: in the final worn configuration, the contact area increased up to five times its initial value and the half contact width reached a value of about $w/2$. The fact that the cut boundaries were properly chosen was demonstrated by the procedure validation results that will be discussed in what follows. In particular, it was demonstrated that the BCs extracted at the selected interfaces did not vary significantly with wear. As an example, Fig. 13 shows the variation of the nodal stress distribution at the DO and DV with the sliding distance d obtained simulating the wear test with the entire global model. A maximum percentage difference on normal stresses of about 20% (the variation of the displacement field was less marked) proved to have an insignificant effect on the wear results.

4.2.2.3. Definition of the boundary conditions

Among the different BC configurations tested, the hybrid approach where nodal forces and displacements were combined and applied on the local model interfaces (Fig. 11) proved to be the most efficient solution. This strategy helps preventing numerical issues due to displacement constraints and, on the other hand, the nodal force constraint ensures contact between the two bodies during the entire wear simulations.

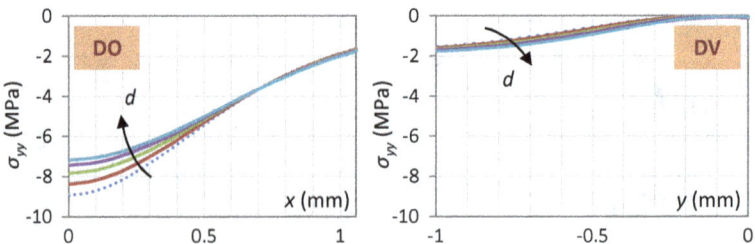

Fig. 13: Trend of the stresses σ_{yy} along the DO and DV boundaries during the wear simulation performed with the global model. (Reprinted from [3] with permission from Elsevier.)

4.2.2.4. Convergence analysis of the local model

The mesh size of the local model was selected based on a typical mesh convergence analysis aimed at reducing the discretisation error on the maximum contact pressure in the initial unworn configuration. An element edge size of about $h_L = a/10$ at the contact region was considered acceptable to accurately predict both contact pressure and wear from the first wear cycles. However, interesting results regarding the effect of mesh size on the accuracy of the wear estimation demonstrated that wear computation is less sensitive to the discretisation error when the contact pressure profile flattens due to the volume loss and the contact becomes more conformal. In this case, a coarse mesh model (i.e. $h_L = a/2$), that roughly predicts wear and stresses in the initial phase, proved to be enough to estimate volume loss at the end of the test with a good level of accuracy. In Fig. 14, the trend of the relative percentage errors on the volume lost (e_A) between the numerical and the theoretical Archard solution as the sliding distance d increases is presented for both fine and coarse local models. It can be noted that this quantity decreased markedly with d and an error of about 10% at the beginning can be reduced to 1% after about $d = 600\,\mathrm{m}$ for the coarse mesh model. A significant reduction of the computational time was observed: the FE wear analysis performed with the coarse mesh model took almost $1/3$ of the time needed by the fine mesh model to complete the simulation. The compromise between high predictive accuracy and low computational cost should be thus evaluated also based on the specific phase of interest of the wear test.

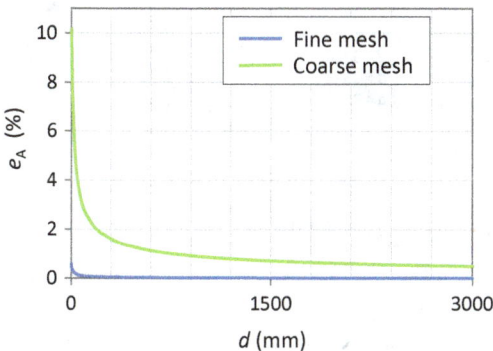

Fig. 14: Trend of the relative percentage errors on the volume lost (e_A) between the numerical and the theoretical Archard solution as the sliding distance d increases for both fine and coarse mesh local models. (Adapted and reprinted from [3] with permission from Elsevier.)

4.2.2.5. Wear results and procedure validation

The main contact and wear results obtained with the local model are shown in Fig. 15. Typical trends of the maximum contact pressure (P_{max}), maximum wear depth (h_{max}) and wear volume with increasing values of the travelled distance d can be observed (Figs. 15(b) and 15(d)). As the wear proceeded, the contact became more conformal due to the volume loss on the pin surface and the contact profile rapidly flattened, as also demonstrated by the evolution of contact pressure and pin wear profiles (Figs. 15(a) and 15(c)). A first running-in phase where the maximum contact pressure drastically decreased and the value of the maximum wear increased rapidly and nonlinearly is followed by a quasi-steady state condition characterised by a linear increase of h_{max} and an almost constant value of P_{max}. These results were in fully agreement with the experimental and numerical study reported in [13] and, more importantly in the context of procedure validation, they were almost identical (discrepancies lower than 0.5%)

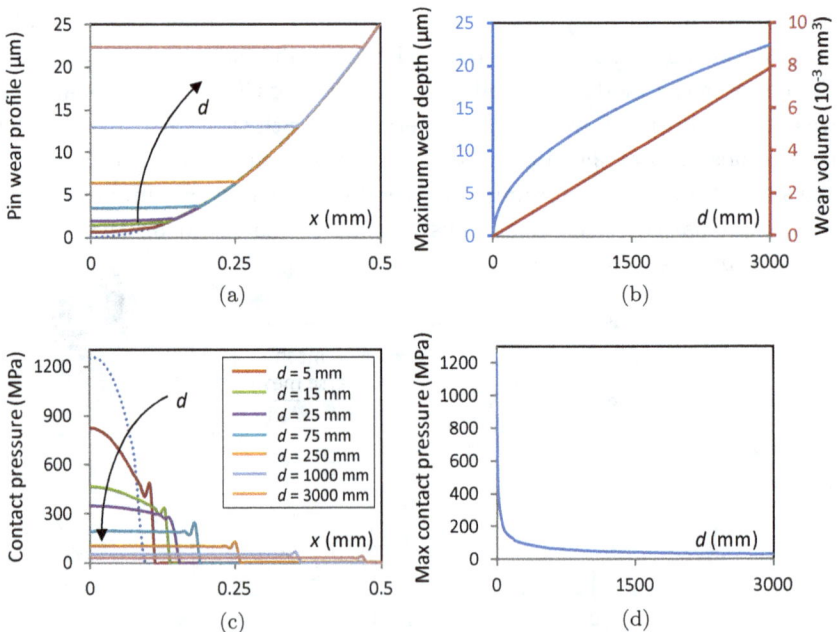

Fig. 15:   Main contact and wear results obtained with the local model in terms of (a) pin wear profile, (b) maximum wear depth and wear volume, (c) contact pressure profile and (d) maximum contact pressure evolution with the sliding distance d. Contact pressure and wear profiles before wear are plotted with dashed lines. (Adapted and reprinted from [3] with permission from Elsevier.)

to the ones obtained simulating the same wear test with the global model. A significant reduction of the computational time was observed using submodelling wear procedure: to reach the same level of accuracy, the global model took more than twice the time needed for the local wear model to complete the simulation.

5. Conclusions

FE modelling allows to fully predict the contact conditions in presence of wear capturing the geometrical changes and the contact pressure evolution, aspects not considered in analytical wear simulations. However, that typically implies high computational costs that can be reduced adopting one or a combination of computational strategies, depending on the simulated case. In particular, in this chapter, the accelerated wear procedure and the submodelling technique are presented and discussed in detail for simple case studies, demonstrating their ability to reduce the simulation duration. However, in order to obtain reliable results, the procedure parameters (e.g. $\bar{N}$ or the cut boundaries geometry) must be set carefully, by means of ad hoc sensitivity analyses. Moreover, as for the analytical wear models, the FE wear model validity should be assessed considering both wear volume and wear map.

References

[1] J. Lengiewicz and S. Stupkiewicz, Efficient model of evolution of wear in quasi-steady-state sliding contacts, *Wear.* **303**, 611–621 (2013).

[2] I. R. McColl, J. Ding, and S. B. Leen, Finite element simulation and experimental validation of fretting wear, *Wear.* **256**, 1114–1127 (2004).

[3] C. Curreli, F. Di Puccio, and L. Mattei, Application of the finite element submodeling technique in a single point contact and wear problem, *Inter J Numer Meth Eng.* **116**, 708–722 (2018).

[4] L. Mattei and F. Di Puccio, Influence of the wear partition factor on wear evolution modelling of sliding surfaces, *Int J Mech Sci.* **99**, 72–88 (2015).

[5] C. Mary and S. Fouvry, Numerical prediction of fretting contact durability using energy wear approach: optimisation of finite-element model, *Wear.* **263**, 444–450 (2007).

[6] L. Johansson, Numerical simulation of contact pressure evolution in fretting, *J Tribol.* **116**, 247–254 (1994).

[7] N. Stromberg, An augmented Lagrangian method for fretting problems, *Eur. J Mech. A/Solids.* **16**, 573–593 (1997).

[8] L. Rodríguez-Tembleque, R. Abascal, and M. H. Aliabadi, A boundary element formulation for wear modeling on 3D contact and rolling-contact problems, *Int J Solids Struct.* **47**, 2600–2612 (2010).

[9] S. Mukras, N. H. Kim, W. G. Sawyer, D. B. Jackson, and L. W. Bergquist, Numerical integration schemes and parallel computation for wear prediction using finite element method, *Wear.* **266**, 822–831 (2009).

[10] N. H. Kim, D. Won, D. Burris, B. Holtkamp, G. R. Gessel, P. Swanson, and W. G. Sawyer, Finite element analysis and experiments of metal/metal wear in oscillatory contacts, *Wear.* **258**, 1787–1793 (2005).

[11] N. G. Cormier, B. S. Smallwood, G. B. Sinclair, and G. Meda, Aggressive submodelling of stress concentrations, *Inter J Numer Meth Eng.* **46**, 889–909 (1999).

[12] C. Curreli, M. Viceconti, and F. D. Puccio, Submodeling in wear predictive finite element models with multipoint contacts, *Inter J Numer Meth Eng.* **122**, 3812–3823 (2021).

[13] P. Põdra and S. Andersson, Simulating sliding wear with finite element method, *Tribol Int.* **32**, 71–81 (1999).

Chapter 7

Modelling Wear Using the Finite Element Method in Abaqus®

I. Khader

Department of Industrial Engineering, German Jordanian University,
P.O. Box 35247, Amman 11180, Jordan
iyas.khader@gju.edu.jo

This chapter presents a computational framework for modelling wear in the finite element method (FEM) using the commercial software package Abaqus®. Abaqus/Standard provides the possibility to model wear based on the arbitrary Lagrangian–Eulerian (ALE) adaptive meshing using the user subroutine UMESHMOTION.

Initially, a friction model must be adopted to lay the foundation for the mechanical interactions; consequently, the adaptive mesh domain must be carefully defined based on the problem under consideration. The implementation of wear in a finite element (FE) framework requires the definition of a wear coefficient that is ideally obtained from tribological tests and a wear equation specifically tailored to adapt to the tribological system being modelled.

The chapter details the steps required to create a functioning wear simulation model. Moreover, it includes some practical aspects to be taken into consideration in modelling systems and interpreting the results.

1. Introduction

Wear has been long recognised as a life-limiting failure mechanism of engineering materials and machine components. The numerical modelling of wear has gained the interest of many researchers as the notion of achieving a quantitative description of wear based on computer simulations would provide engineers with invaluable information in designing tribological systems.

Since wear in a tribological system is an intrinsically complex problem that depends on the loading, contact mechanics and materials behaviour in addition to many other aspects, its incorporation in a finite element method (FEM) framework would allow coupling several phenomena and taking advantage of the vast capabilities of this versatile numerical method.

Although finite element (FE) simulations are extensively applied in failure prediction of engineering systems, which diminishes the need to conduct costly and time-consuming testing programs, the concept is somewhat different when considering numerical simulations of wear. Since there is no universal equation that can be used to describe wear, the same applies to friction as well, and since wear is system-specific, wear rates generated in one tribological system can hardly be adopted to other systems or even system configurations. From what preceded, it should be kept in mind that the simulation of wear requires conducting indispensable tribological tests, from which wear rates are obtained. These wear rates may then be converted into wear coefficients that find their way as input data in numerical simulations. The purpose of wear simulations is thus to provide the designer with information pertaining to stresses, strains and temperature fields as a function of evolving geometrical modifications that occur due to progressive wear. Such information is generally very difficult, if not impossible, to obtain from experiments or even computations based on closed form solutions.

For instance, during operations, changes occurring to the geometry of a part may alter the contact area and consequently, the contact stresses, thus, resulting in fundamental changes to the wear mechanism itself. Therefore, in many cases, judging a tribological system based on its initial contact configuration (i.e. initial contact area, contact pressure, etc.) may lead to false assumptions resulting in poor design considerations.

In this chapter, the implementation of wear simulations in the commercial software package Abaqus®[1] will be detailed and some examples will be given to provide a systematic procedure to create a functional simulation.

2. General Procedure

Abaqus allows the simulation of wear through the concept of adaptive meshing and adaptive mesh constraints to accommodate for nodal ablation in Abaqus/Standard. Adaptive meshing encompasses two concepts, namely, pure Lagrangian and pure Eulerian. In a Lagrangian analysis the material

---

[1] Abaqus and SIMULIA are registered trademarks of Dassault Systèmes or its affiliates.

does not leave the mesh or, in other words, nodes are fixed within the material and elements are completely filled with a single material. Simply put, the material boundary coincides with an element boundary. On the other hand, in an Eulerian analysis nodes are fixed in space and material flows through elements that do not deform. Hence, Eulerian elements may not always be completely full of material and may be partially or completely void. The Eulerian material boundary does not correspond to an element boundary and the material may leave the mesh and be removed during a simulation. For the simulation of wear, Abaqus/Standard uses an arbitrary Lagrangian–Eulerian (ALE) analysis, in which the mesh is allowed to move independently of the material and the material is removed at the boundaries to model the effects of ablation. ALE adaptive meshing does not alter elements and nodes' connectivity and it is extremely important to realise that it does not allow the creation or destruction of elements. This poses limitations on its ability to maintain a high-quality mesh especially in cases of extreme deformation levels.

To work with ALE adaptive meshing, both an adaptive mesh domain and an adaptive mesh constraint must be defined. To model material loss to ablation, the adaptive mesh constraint may be thought of as a wear coefficient and is defined in terms of an ablation velocity. This coefficient and the general solution of ablation are handled by describing the wear equation in the user subroutine UMESHMOTION, which is available for Abaqus/Standard.

Throughout the simulation, the user subroutine UMESHMOTION is called at the end of any increment where adaptive meshing is performed. It is called for any given node for every mesh smoothing sweep, whose frequency may be set by the user. Mesh velocities computed by the Abaqus/Standard meshing algorithm for that node are passed into UMESHMOTION, which modifies them to account for the ablation velocity computed at that node. The modified velocity is determined according to the set of equations coded in the subroutine based on various problem-specific variables and parameters. A simplified structure of the simulation is shown in Fig. 1.

3. Friction Models

Before going into the details of a wear simulation, a general understanding of the friction models provided in Abaqus is imperative. It is important to note that this section is not meant to substitute or comprehensively address

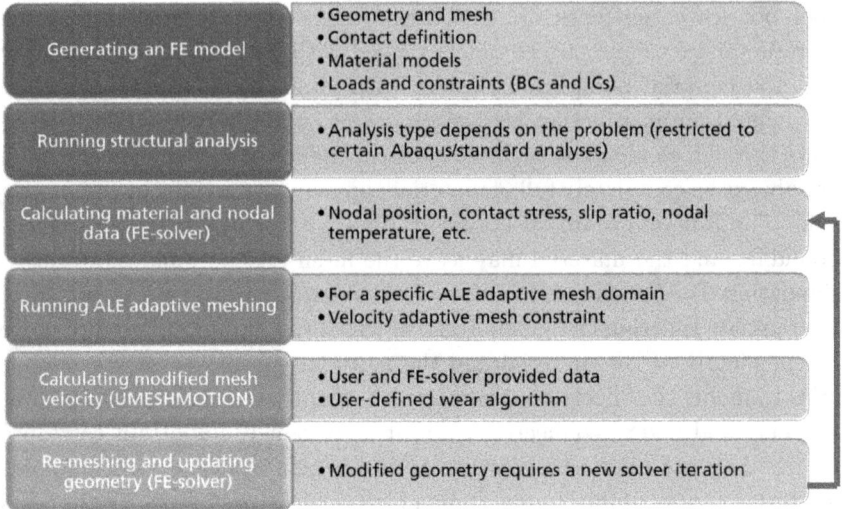

Fig. 1:   General wear simulation sequence in Abaqus.

the modelling of frictional behaviour in Abaqus as this may be found in the Abaqus Interactions Guide [1]. Therefore, we will limit our discussion to the most common cases supported in Abaqus/Standard.

Friction occurs due to several mechanisms that may act together depending on various circumstances. Hence, there is no generalised mathematical model that could accurately describe friction in a tribological system [2]. To overcome this complexity, modelling friction is usually accomplished by transmitting shear stresses and normal stresses across the interface of contacting bodies in order to obtain a quantitative description of friction forces. In addition to user-defined friction models provided by the subroutines FRIC and FRIC_COEF, Abaqus/Standard allows defining friction by means of the classical isotropic Coulomb friction model in addition to its extensions. The Coulomb friction model defines the maximum allowable shear stress across an interface as a function of the normal contact pressure between the contacting surfaces. Hence, the two contacting surfaces are capable of carrying shear stresses across their interface up to a certain magnitude (τ_{crit}), beyond which slipping occurs. The region within the contact area, in which no relative sliding occurs, will be in a "sticking" condition. Generally, if the materials of two contacting solids are dissimilar, the tangential displacements will be different, thus, giving rise to interfacial slip. Friction will oppose this slip and may even prevent it altogether.

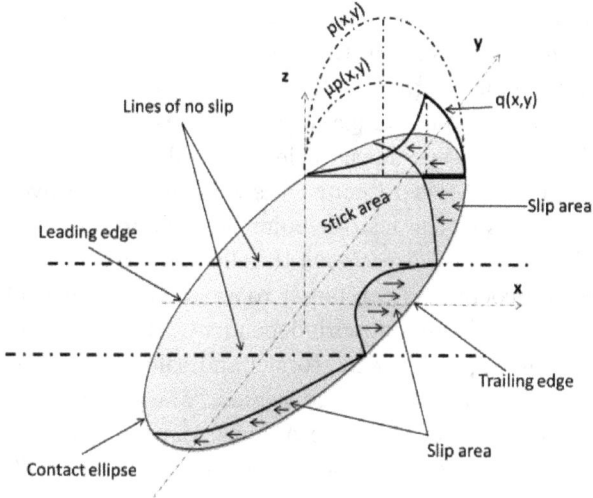

Fig. 2: Microslip zones in a deep-groove ball bearing; the normal contact pressure is denoted by $p(x, y)$, the shear stress is denoted by $q(x, y)$ and the friction coefficient is μ. Adopted from Halme and Andersson [4].

Hence, within the contact zone, some regions of the surfaces will be sticking together while other regions, usually towards the edges of the contact, will undergo slip [3]; this is demonstrated in the contact ellipse of a deep-groove ball bearing in Fig. 2.

3.1. Basic Coulomb friction model

Defining friction in Abaqus may be accomplished by using the keyword *FRICTION, which must be used in conjunction with a pre-defined surface interaction (keyword: *SURFACE INTERACTION).

Abaqus assumes no relative motion between the surfaces if the equivalent frictional stress τ_{eq} remains below τ_{crit}. In a 3D simulation, τ_{eq} combines the two orthogonal shear stress components acting in the local tangential directions of the contact surfaces or contact elements as follows:

$$\tau_{eq} = \sqrt{\tau_1^2 + \tau_2^2}, \tag{1}$$

whereas the critical shear stress proportional to the normal contact pressure (p) is given by

$$\tau_{crit} = \mu . p, \tag{2}$$

where the friction coefficient $\mu = \mu(\dot{\gamma}_{eq}, p, \bar{\theta}, \bar{f})$ may be defined as a function of the contact pressure p, the equivalent slip rate (or slip velocity) $\dot{\gamma}_{eq} = \sqrt{\dot{\gamma}_1^2 + \dot{\gamma}_2^2}$, which is given in terms of the two orthogonal slip velocity components $\dot{\gamma}_1$ and $\dot{\gamma}_2$, the average surface temperature at the contact point $\bar{\theta} = \frac{1}{2}(\theta_A + \theta_B)$ and the average user-defined field variable(s) at the contact point $\bar{f} = \frac{1}{2}(f_A + f_B)$, where point A is a node on the slave surface and point B corresponds to the nearest point on the opposing master surface. If friction is isotropic, the direction of the slip and the frictional stress coincide. The friction coefficient can be set to any non-negative value; if set to $\mu > 0.2$ or defined as being pressure-dependent, Abaqus will automatically invoke an "unsymmetric" matrix storage and solution scheme to improve convergence rate as explained in the section "Matrix storage and solution scheme in Abaqus/Standard" in the Abaqus Analysis Guide [5]. The basic model of friction assumes isotropic friction and thus, a constant μ in all directions.

The condition of no relative motion in the absence of slip is modelled in Abaqus by either

1. an approximated stiff elastic behaviour (i.e. default penalty method), by which the relative motion from the position of zero shear stress is bounded by the value of the maximum allowable elastic slip (γ_{crit}), or
2. set exactly to zero through the adoption of a Lagrange multiplier formulation (invoked by using the option LAGRANGE in the friction definition). Although capable of enforcing the exact and more accurate sticking condition, Lagrange multipliers increase the cost of analysis due to the addition of degrees of freedom to the model and an increase in the number of iterations needed to achieve convergence. Moreover, this formulation may slow or even prevent convergence, especially in models with many points undergoing a change between sticking and slipping conditions.

The condition of no relative motion may also be enforced through a special case of friction, called *rough friction* (option ROUGH), which assumes an infinite value for μ and no limit to the shear stress at the interface and thus, no relative motion as long as the surfaces remain in contact. Rough friction does not require a value for the coefficient of friction (μ) and is internally implemented by Abaqus using the Lagrange multiplier formulation without any additional input from the user. It is generally intended for cases of non-intermittent contact (i.e. once the surfaces close, they must remain closed throughout the simulation).

3.1.1. *Applying the basic friction model*

The basic friction model may be applied by using the keyword *FRICTION. If not followed by the option DEPENDENCIES, it is assumed that the friction coefficient has no field variable dependencies and is thus, dependent only on the slip rate, contact pressure and temperature. The most robust method to apply this model is the approximated stiff elastic behaviour, which may be set by one of the following three methods:

1. Defining an absolute magnitude for the allowable elastic slip (γ_{crit}). This is accomplished by using the option ELASTIC SLIP in the friction definition as follows:

   `*FRICTION, ELASTIC SLIP=`γ_{crit}

2. Invoking the default value of the slip tolerance (F_f), which is defined as the maximum allowable elastic slip to the characteristic contact surface face dimension. This default value is equal to 0.005 of the average length of all contact elements in the model. This may be set by specifying no options in the definition of friction, i.e.

   `*FRICTION`

3. Overriding the default value of the slip tolerance (F_f) by adopting a value larger than 0.005 to increase computational efficiency or a smaller value to improve accuracy. This may be accomplished by using the option SLIP TOLERANCE as follows:

   `*FRICTION, SLIP TOLERANCE=`F_f

   It should be noted that in the case of steady-state transport analysis the elastic slip rate substitutes the elastic slip in the above cases.

3.1.2. *Defining static and kinetic friction coefficients using the basic friction model*

In the basic Coulomb friction model, it is also possible to specify a static friction coefficient (μ_s) and a kinetic friction coefficient (μ_k), the latter of which is sometimes referred to as the dynamic friction coefficient. Experimental data have shown that the static friction coefficient is higher than its kinetic counterpart.

The default model requires the static friction coefficient to be given at a slip rate equal to zero, while defining the kinetic friction coefficient at the highest slip rate. In this case, both the static and kinetic friction coefficients can be functions of the contact pressure, temperature and field variables. It can be applied as follows:

```
*FRICTION
μₛ, 0
μₜ₁, γ̇_eq,1
μₜ₂, γ̇_eq,2
...
μₜₙ, γ̇_eq,n
μₖ, γ̇_eq,max
```

where the values μ_{t1} and $\mu_{t2} \ldots \mu_{tn}$ correspond to their respective slip rates $\dot{\gamma}_{eq,1}, \dot{\gamma}_{eq,2} \ldots \dot{\gamma}_{eq,n}$ and define the transitional values of the friction coefficient between μ_s at $\dot{\gamma}_{eq} = 0$ and μ_k at $\dot{\gamma}_{eq,max}$.

3.1.3. *Defining static and kinetic friction coefficients using exponential decay*

In addition to the basic model described above, Abaqus provides a model that explicitly defines the transition between μ_s and μ_k in terms of an exponential decay (option EXPONENTIAL DECAY) by directly specifying the decay coefficient (d_c) according to the equation [6–8]

$$\mu = \mu_k + (\mu_s - \mu_k) \exp -d_c \dot{\gamma}_{eq}, \tag{3}$$

where the decay coefficient has a default value of $d_c = 0$ if left unspecified. Moreover, the use of test data to fit the exponential model is possible (option EXPONENTIAL DECAY, TEST DATA). In this case, the user must provide at least two data points; the first of which represents the static coefficient of friction specified at $\dot{\gamma}_{eq} = 0$ and the second at a measured finite slip rate. Both models that rely on an exponential decay function in defining the transition between μ_s and μ_k may be used only with isotropic friction and do not allow dependence on contact pressure, temperature or field variables. Further details on these models may be found under section "Specifying static and kinetic friction coefficients" in [1].

3.1.4. *Defining a shear stress limit*

By using the option TAUMAX in the definition of friction, it is possible to enforce a limit on the critical stress, thus allowing slip to occur once

τ_{eq} becomes equal to or larger than either the user-defined equivalent shear stress limit (τ_{max}) or the critical stress value defined by Eq. (2), whichever is smaller, hence

$$\tau_{crit} = \min\left(\mu p, \tau_{max}\right). \tag{4}$$

This model is especially useful in cases where the contact pressure is expected to be extremely high such as in metal forming processes.

3.1.5. *Applying anisotropic friction*

This model may be invoked by using the option ANISOTROPIC in the definition of friction; it allows defining different friction coefficients for the two (principal) orthogonal directions on the contact surface. The definition of the anisotropic model, hence, requires specifying two friction coefficients μ_1 and μ_2 for the first and second local tangent directions, respectively. In this model, the direction of slip is assumed to be orthogonal to the critical shear stress surface τ_{crit}, which forms an ellipse in the shear space ($\tau_1 - \tau_2$). This ellipse intersects the τ_1 axis at point $\mu_1 p$ and the τ_2 axis at point $\mu_2 p$, whereby

$$\left(\frac{\tau_1}{\mu_1}\right)^2 + \left(\frac{\tau_2}{\mu_2}\right)^2 = p^2. \tag{5}$$

Similar to the isotropic friction model, the anisotropic model can depend on slip rate, contact pressure, temperature and user-defined field variables.

4. Adaptive Meshing

In this section, we will go through the adaptive meshing procedure required to model incremental wear in Abaqus/Standard. The following two terms are important for the discussion:

1. Adaptive mesh domain: the part of the FE mesh that is defined to undergo wear.
2. Adaptive mesh constraints: the constraints that define the motion of nodes in an adaptive mesh domain. Adaptive mesh constraints may either define mesh motions that are independent of the underlying material for a set of nodes (spatial constrains) or nodes that must follow the material (i.e. Lagrangian constraints).

The ablation (wear) of material in Abaqus/Standard is modelled by defining an adaptive mesh domain and by prescribing adaptive mesh motion constraints. To model wear, the adaptive mesh domain will be remeshed to account for material recession by moving material points and nodal points to new locations. For the sake of consistency, it will be assumed that the ablation equation describes the material recession rate normal to the surface in terms of a mesh velocity that is defined as a local function of solution quantities (e.g. stress, temperature, sliding distance, etc.). This equation will be defined in the UMESHMOTION subroutine as will be detailed in the next section. The solver enables imposing this velocity and maintaining it as mesh constraint as the surface moves. Consequently, subsurface nodes will be adjusted to account for material loss.

4.1. *Defining an adaptive mesh domain*

The adaptive mesh domain is defined by using the keyword *ADAPTIVE MESH. An element set that contains all the solid elements in the adaptive mesh domain must be defined.

The frequency and intensity of adaptive meshing for that domain can be defined by setting the frequency in increments at which adaptive meshing is to be performed (option FREQUENCY) and the number of mesh sweeps to be performed in each adaptive mesh increment (option MESH SWEEPS).

4.2. *Defining adaptive mesh constraints*

Once an adaptive mesh domain has been defined, mesh constraints corresponding to this domain must be carefully prescribed. This is accomplished by using the keyword *ADAPTIVE MESH CONSTRAINT.

As previously mentioned, two types of adaptive mesh constraints may be defined: spatial or Lagrangian; and two type of motions may be prescribed: displacement or velocity. Simply put, to model material ablation, spatial adaptive mesh constraints must be prescribed to the domain. Whether these spatial constraints are defined as velocity or displacement strictly depends on the ablation equation defined in UMESHMOTION. However, to be able to define incremental wear, velocity constraints must be defined. Nevertheless, prescribing Lagrangian adaptive mesh constraints remains essential in defining the boundaries of the adaptive mesh domain, whereby the nodes on the boundary of the adaptive mesh domain where it meets the regular mesh must be considered as Lagrangian. Consequently, the model will include an adaptive mesh domain that undergoes wear adjacent to a non-adaptive mesh, both separated by Lagrangian adaptive mesh constraints.

4.3. *Defining adaptive mesh controls*

Various aspects of the adaptive meshing and advection algorithms applied to an adaptive mesh domain may be controlled by using the keyword *ADAPTIVE MESH CONTROLS after having set the option CONTROLS in *ADAPTIVE MESH equal to the name of the *ADAPTIVE MESH CONTROLS associated with that particular domain.

Of all options associated with this keyword, only a few can be used in Abaqus/Standard and even fewer are meaningful to apply in a wear simulation. The MESHING PREDICTOR options may be used in certain cases to either perform adaptive meshing based on the nodal positions at the beginning of the current adaptive mesh increment or based on the default setting of nodal positions in the original mesh.

5. Defining the Ablation Equation

The ablation equation described in this section will define material recession normal to the surface. It will be defined as a mesh velocity that is dependent on specific solution quantities defined by the user (e.g. stress, temperature, sliding distance, etc.). The equation is defined in the subroutine UMESH-MOTION, which should be coded as a standalone computer program in FORTRAN. This FORTRAN script must be saved as an individual file preferably with the extension f.

5.1. *UMESHMOTION structure*

The basic UMESHMOTION subroutine structure, as defined in the Abaqus User Subroutines Guide [9], should be as follows:

```
      SUBROUTINE UMESHMOTION(UREF,ULOCAL,NODE,NNDOF,
     $     LNODETYPE,ALOCAL,NDIM,TIME,DTIME,PNEWDT,
     $     KSTEP,KINC,KMESHSWEEP,JMATYP,JGVBLOCK,LSMOOTH)
C
      INCLUDE 'ABA_PARAM.INC'
C
      DIMENSION ULOCAL(NDIM),JELEMLIST(*)
      DIMENSION ALOCAL(NDIM,*),TIME(2)
      DIMENSION JMATYP(*),JGVBLOCK(*)
C
      user code to define ULOCAL
      and optionally PNEWDT
```

```
C
      RETURN
      END
```

5.1.1. *UMESHMOTION variables*

As in the case of any Abaqus subroutine, the associated variables are divided into three types: (i) *variables to be defined,* (ii) *variables that can be updated* and (iii) *variables passed in for information.* The first type is basically the numerical outcome of the subroutine and thus, must be defined by the user. The second type of variables may be optionally defined if deemed necessary by the user. On the other hand, the *variables passed for information* should never be tampered with. It is also strongly recommended not to change the above shown structure to ensure proper functionality of the subroutine.

In the user subroutine UMESHMOTION, the only *variable to be defined* is ULOCAL; hence, this variable must be defined by the user since its value will be returned to Abaqus to accordingly modify the mesh at the end of each adaptive mesh increment or as specified by the user in the *ADAPTIVE MESH definition.

The variables associated with UMESHMOTION are explained in detail in the Abaqus User Subroutines Guide [9]; for convenience, they will be reproduced in what follows without modification. At this juncture, the reader is strongly recommended to go through all variables and understand their function within the code.

Variable to be defined

ULOCAL
Components of the mesh displacement or velocity of the adaptive mesh constraint node, described in the coordinate system ALOCAL. ULOCAL will be passed into the routine as values determined by the mesh smoothing algorithm. All components of the mesh displacement or velocity will be applied; i.e. you do not have the ability to select the directions in which the mesh displacement should be applied.

Variables that can be updated

PNEWDT
Ratio of suggested new time increment to the time increment currently being used (DTIME, see in what follows). This variable allows you to provide

input to the automatic time incrementation algorithms in Abaqus/Standard (if automatic time incrementation is chosen).

PNEWDT is set to a large value before each call to UMESHMOTION.

The suggested new time increment provided to the automatic time integration algorithms is PNEWDT × DTIME, where the PNEWDT used is the minimum value for all calls to user subroutines that allow redefinition of PNEWDT for this increment.

If automatic time incrementation is not selected in the analysis procedure, values of PNEWDT greater than 1.0 will be ignored and values of PNEWDT less than 1.0 will cause the job to terminate.

LSMOOTH

Flag specifying that surface smoothing be applied after application of the mesh motion constraint. Set LSMOOTH to 1 to enable surface smoothing. When this flag is set, the constraint defined in ULOCAL will be modified by the smoothing algorithm. In cases where ULOCAL describes mesh motion normal to a surface, the smoothing will have a minor impact on this normal component of mesh motion.

Variables passed in for information

UREF

The value of the user-specified displacement or velocity provided as part of the adaptive mesh constraint definition. This value is updated based on any amplitude definitions used with the adaptive mesh constraint or default ramp amplitude variations associated with the current step.

NODE

Node number.

NNDOF

Number of degrees of freedom at the node.

LNODETYPE

Node type flag.

LNODETYPE=1 indicates that the node is on the interior of the adaptive mesh region.

LNODETYPE=2 indicates that the node is involved in a tied constraint.

LNODETYPE=3 indicates that the node is at the corner of the boundary of an adaptive mesh region.

LNODETYPE=4 indicates that the node lies on the edge of a boundary of an adaptive mesh region.

LNODETYPE=5 indicates that the node lies on a flat surface on a boundary of the adaptive mesh region.

LNODETYPE=6 indicates that the node participates in a constraint (other than a tied constraint) as a master node.

LNODETYPE=7 indicates that the node participates in a constraint (other than a tied constraint) as a slave node.

LNODETYPE=10 indicates that a concentrated load is applied to the node.

ALOCAL
Local coordinate system aligned with the tangent to the adaptive mesh domain at the node. If the node is on the interior of the adaptive mesh domain, ALOCAL will be set to the identity matrix. In other cases the 1-direction is along an edge or in the plane of a flat surface. When NDIM=2, the 2-direction is normal to the surface. When NDIM=3, the 2-direction also lies in the plane of a flat surface or is arbitrary if the node is on an edge. When NDIM=3, the 3-direction is normal to the surface or is arbitrary if the node is on an edge.

NDIM
Number of coordinate dimensions.

TIME(1)
Current value of step time.

TIME(2)
Current value of total time.

DTIME
Time increment.

KSTEP
Step number.

KINC
Increment number.

KMESHSWEEP
Mesh sweep number.

JMATYP
Variable that must be passed into the GETVRMAVGATNODE utility routine to access local results at the node.

JGVBLOCK

Variable that must be passed into the GETVRN, GETNODETOELEMCONN and GETVRMAVGATNODE utility routines to access local results at the node.

5.1.2. *Utility routines*

Utility routines are predefined routines that can be "called" from within a user subroutine (e.g. UMESHMOTION) to perform specific tasks such as accessing data. Each subroutine has a list of associated utility routines with which they are usable. Within UMESHMOTION, the following utility routines may be called:

- GETVRN: to access node point information (reference: Utility Routines/Obtaining node point information [9]).
- GETNODETOELEMCONN: to retrieve a list of elements connected to a specified node (reference: Utility Routines/Obtaining node to element connectivity [9]).
- GETVRMAVGATNODE: to access material integration point information averaged at a node and is thus restricted to real-valued results (reference: Utility Routines/Obtaining material point information averaged at a node [9]).

Each utility subroutine has its associated variables, which are divided into two types: (i) *variables to be provided to the utility routine* and (ii) *variables returned from the utility routine.*

To fully understand the functionality and limitations of each utility routine, the reader is encouraged to refer to its detailed description in the Utility Routines Section of the Abaqus User Subroutines Guide [9].

5.2. *Example 1: Wiring a UMESHMOTION subroutine to simulate the wear of material as a function of the contact pressure*

In this example, a material will be ablated based on the nodal contact pressure. To limit our focus to the structure of UMESHMOTION, only the subroutine will be presented here.

```
1        SUBROUTINE UMESHMOTION(UREF,ULOCAL,NODE,NNDOF,
2     $      LNODETYPE,ALOCAL,NDIM,TIME,DTIME,PNEWDT,
3     $      KSTEP,KINC,KMESHSWEEP,JMATYP,JGVBLOCK,LSMOOTH)
```

```
4    C
5            INCLUDE 'ABA_PARAM.INC'
6    C
7            CHARACTER*80 PARTNAME
8            PARAMETER (NELEMMAX=100)
9            DIMENSION ULOCAL(NDIM)
10           DIMENSION ALOCAL(NDIM,*)
11           DIMENSION JGVBLOCK(*), JMATYP(*)
12           DIMENSION ARRAY(600)
13           DIMENSION JELEMLIST(NELEMMAX),JELEMTYPE(NELEMMAX)
14           REAL XCPRESS
15           REAL SURFV
16   C
17           NELEMS=NELEMMAX
18           JTYP=0
19   C
20           CALL GETNODETOELEMCONN(NODE,NELEMS,JELEMLIST,JELEMTYPE,
21        $      JRCD,JGVBLOCK)
22   C
23           LOCNUM=0
24           JRCD=0
25           PARTNAME='   '
26   C
27           CALL GETPARTINFO(NODE,JRCD,PARTNAME,LOCNUM,JRCD)
28   C
29           CALL GETVRMAVGATNODE(LOCNUM,JTYP,'CSTRESS',ARRAY,JRCD,
30        $      JELEMLIST,NELEMS,JMATYP,JGVBLOCK)
31           XCPRESS=ARRAY(1)
32   C
33           SURFV=XCPRESS*UREF
34   C
35   C        Applying wear
36   C
37           DO KDIM=1, NDIM
38             DO JDIM=1, ndim
39                ULOCAL(KDIM)=ULOCAL(KDIM)-ALOCAL(JDIM,KDIM)*SURFV
40             ENDDO
41           ENDDO
42   C
43           RETURN
44           END
```

Declaring variables (lines 7–15)

In FORTRAN it is essential to declare (i.e. define) all variables used in the program to avoid unexpected outcomes. The two intrinsic data types used in this example script are CHARACTER and REAL. Fixed parameters and multi-dimensional arrays were also declared by using the keywords PARAMETER and DIMENSION, respectively.

This example subroutine limits the number of elements in the model to 100. This is given in the definition of the parameter NELEMMAX=100.

It is recommended to consult the variable name rules in FORTRAN to avoid using unacceptable or reserved names, which may otherwise result in unexpected results. It is also recommended to consult version-specific syntaxes such as comments, continuation lines, etc.

Retrieving model information (line 27)

The utility subroutine GETPARTINFO was called to retrieve model information from Abaqus. This utility subroutine may be called from within any Abaqus subroutines. It retrieves the part instance name and original node or element number corresponding to an internal node or element number. It is called using the following syntax:

```
CALL GETPARTINFO(NODE,JRCD,PARTNAME,LOCNUM,JRCD)
```

Obtaining contact pressure (lines 29–30)

In order to obtain contact stress values such as CPRESS (contact pressure), one may call the utility subroutine GETVRMAVGATNODE using the appropriate *output variable identifier*. The syntax is as follows:

```
CALL GETVRMAVGATNODE(NODE,JTYP,'VAR',ARRAY,JRCD,
$      JELEMLIST,NELEMS,JMATYP,JGVBLOCK)
```

The *variable key* VAR must be substituted by the appropriate "output variable key" that may be found in the Abaqus Output Guide [10]. By searching the section "Abaqus/Standard output variable identifiers/Surface variables" of the Abaqus Output Guide [10], one finds that CPRESS may be obtained by using the output variable identifier CSTRESS, which upon retrieving by calling the utility routine GETVRMAVGATNODE, returns the following attributes:

- CPRESS (contact pressure between the node on the slave surface and the master surface with which it interacts),

- CSHEAR1 (frictional shear traction component in the local 1-direction on the master surface), and
- CSHEAR2 (frictional shear traction component in the local 2-direction on the master surface for 3D).

The syntax to retrieve CSTRESS by calling GETVRMAVGATNODE is the following

```
CALL GETVRMAVGATNODE(LOCNUM,JTYP,'CSTRESS',ARRAY,JRCD,
$      JELEMLIST,NELEMS,JMATYP,JGVBLOCK)
```

where ARRAY is an arbitrary variable name set by the user to define an array that will store the three aforementioned values (CPRESS, CSHEAR1 and CSHEAR2). As shown in *line 31*, the value of CPRESS was retrieved from the first element of the array variable ARRAY, i.e. ARRAY(1), and then stored into the pre-defined variable XCPRESS.

It is strongly recommended to check the order of returned components for each individual label in the definition of GETVRMAVGATNODE.

Only output variable keys that are valid for results file output are available for use with GETVRMAVGATNODE. This may be verified by ensuring that the desired output variable is available for the *.fil* file format. By going back to the section "Abaqus/Standard output variable identifiers/Surface variables" of the Abaqus Output Guide [10], the description of CSTRESS is given as follows:

CSTRESS

.dat: yes .fil: yes .odb Field: yes .odb History: yes

Contact pressure (CPRESS) and frictional shear stresses (CSHEAR). Output is also available on the master surface to the .odb file in a single master-slave setting.

Hence, CSTRESS is available for the *.fil* file format (.fil: yes), and may be retrieved by calling the utility routine GETVRMAVGATNODE.

Calculating wear and ablating nodes (lines 33–41)

In this example, wear was defined in terms of the nodal ablation velocity SURFV as follows:

```
SURFV=XCPRESS*UREF
```

and hence, it is given in terms of the rate of nodal ablation as follows:

$$\dot{h}_i = p_i U_{\text{ref}}, \tag{6}$$

where p_i is the nodal contact pressure and U_{ref} is the user-defined wear coefficient (constant) defined in the *ADAPTIVE MESH CONSTRAINT definition (not shown here). Since U_{ref} is a constant, the wear in this example is defined as an explicit function of the contact pressure only.

The application of wear on the respective nodes is accomplished as follows:

```
ULOCAL(kdim)=ULOCAL(kdim)-ALOCAL(jdim,kdim)*SURFV
```

whereby the only required output of the subroutine (ULOCAL) is prescribed to each node by subtracting the value of SURFV, in the local coordinate system (ALOCAL) aligned with the tangent to the adaptive mesh domain at the node, from the previously stored value of ULOCAL for the same node.

6. Creating a Wear Simulation

In this section, the procedure for creating a fully functional wear simulation will be detailed. It is assumed that the user possesses enough knowledge to create and successfully run a structural-mechanics contact simulation in Abaqus/Standard. Details like meshing, material modelling, assignment of properties, loads and boundary conditions, contact definition, etc. will not be addressed here.

6.1. *Example 2: Creating a functional model to simulate wear based on the Holm–Archard equation*

This example details how to create a 2D wear simulation based on a pin-on-disk sliding contact experiment as shown in Fig. 3. In this experiment, the pin was made of a ceramic (e.g. silicon nitride) and it was modelled as an elastic material; whereas, the disk was made of a metal (e.g. Inconel 718) with elastic–plastic properties. This model example is based on the work published by Khader *et al.* [11].

Material ablation will be modelled in the pin only and according to the Holm–Archard [13, 14] wear equation

$$\dot{V}_w = K p \dot{s} A, \tag{7}$$

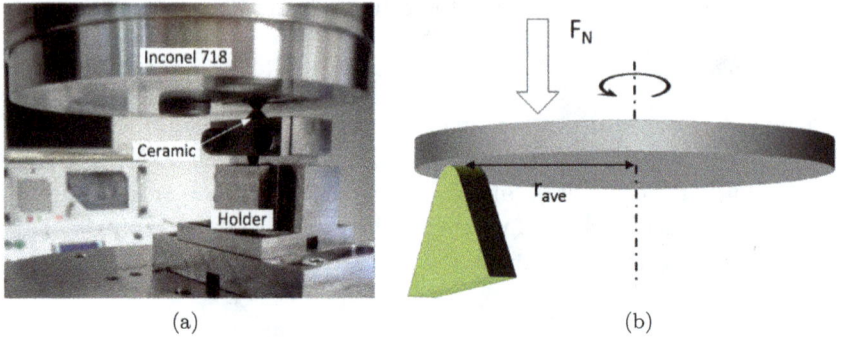

(a)   (b)

Fig. 3: (a) Pin-on-disk sliding contact experimental setup at the Fraunhofer Institute for Mechanics of Materials IWM [12], (b) schematic diagram of the pin-on-disk experiment showing the contact between the ceramic pin and the metallic disk.

where $\dot{V}_w$ is the volumetric wear rate, p is the contact pressure within the contact area A and $\dot{s}$ is the slip rate. The wear coefficient K is given by

$$K = \frac{V_w}{F_N s},\qquad(8)$$

where s is the sliding distance and F_N is the applied normal force. The wear coefficient must be obtained from the tribological tests; preferably from tests carried out under the same contact conditions. Its unit is volume/force · length and in SI units it is mm^3/N · m. Note that the volume (nominator) has the unit mm^3 and the force and sliding distance (denominator) are given in N and m, respectively. This selection of unit stems from practical aspects.

6.1.1. *Geometry and FE-mesh*

The geometry of the problem consists of two 2D bodies modelling a pin (bottom) and a disk (top); see Fig. 4. Both the pin and the disk were discretised with a uniform mesh. Four-node bilinear plane stress elements (CPS4) were chosen for the discretisation, which allows accurate modelling of a linear contact. The dimensions of the disk are $1.0 \times 0.5\,mm^2$, whereas the pin has a height of 60 μm and an arc length of 400 μm; the radius of the pin is 0.8 mm.

6.1.2. *Loads, boundary conditions and contact interactions*

In this example, the disk is fixed from its top nodes and the pin presses and slides against it. To move the pin, multi-point constraints of type

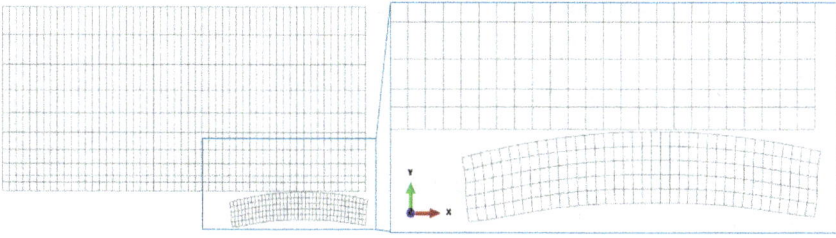

Fig. 4: FE-mesh of the pin-on-disk in Example 2.

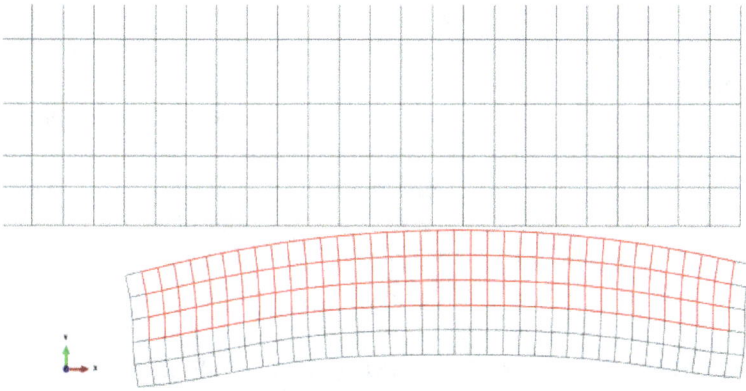

Fig. 5: Adaptive mesh domain defined by the element set adaptive_elem.

EQUATION were prescribed at its bottom nodes. Contact was defined between the pin and disk using the CONTACT PAIR formulation, whereby the pin acts as slave and the disk as master. Note that wear can be implemented only on the slave surface. The contact was initialised by applying a force of 33.3 N/mm in the positive y-direction. Once contact has been initialised, the pin is moved in the negative x-direction at a rate of 1,000 mm/s using velocity boundary conditions.

6.1.3. *Adaptive mesh domain and constraints*

As detailed in Section 4, an adaptive mesh domain requires the definition of its elements, the nodes upon which spatial mesh constraints are applied and its boundaries.

The adaptive mesh domain was set by defining an element set (element set name: adaptive_elems) to include the elements highlighted in Fig. 5.

Setting the adaptive meshing frequency to one, i.e. adaptive mesh will be applied at the end of each increment, and setting the number of mesh sweep per increment to one, the adaptive mesh domain can be defined as follows:

```
*ADAPTIVE MESH, ELSET=adaptive_elem, FREQUENCY=1,
MESH SWEEPS=1, OP=NEW
```

Node ablation is applied by defining spatial-type adaptive mesh constraints on the desired node set (node set name: adaptive_nodes) as shown in Fig. 6. It is important to note that spatial adaptive mesh constraints should not be assigned to nodes shared between the adaptive mesh domain and the regular mesh surrounding it.

Velocity-type spatial constraints were applied on the node set adaptive_nodes. The mesh motion velocity must correspond to the wear coefficient obtained experimentally as will be detailed in the following section. For the sake of simplicity, we will assume a value of $1 \times 10^{-8}\,\mathrm{mm}^3/\mathrm{N}{\cdot}\mathrm{m}$ for the mesh motion velocity. Accordingly, the spatial mesh constraints may be defined as follows:

```
*ADAPTIVE MESH CONSTRAINT, CONSTRAINT TYPE=SPATIAL,
TYPE=VELOCITY, USER
adaptive_nodes,,,1.0e-8
```

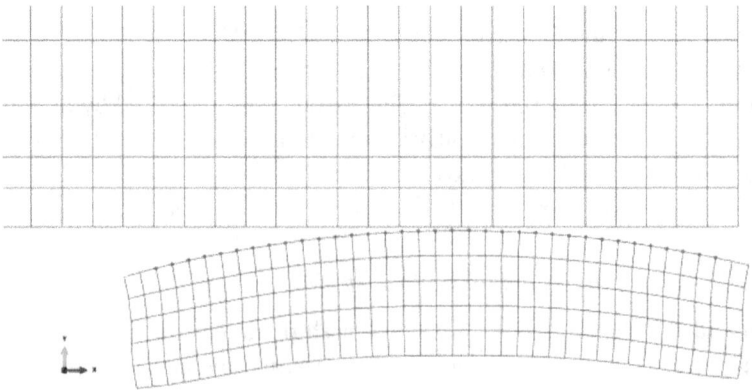

Fig. 6:   Spatial adaptive mesh constraints applied on the node set adaptive_nodes.

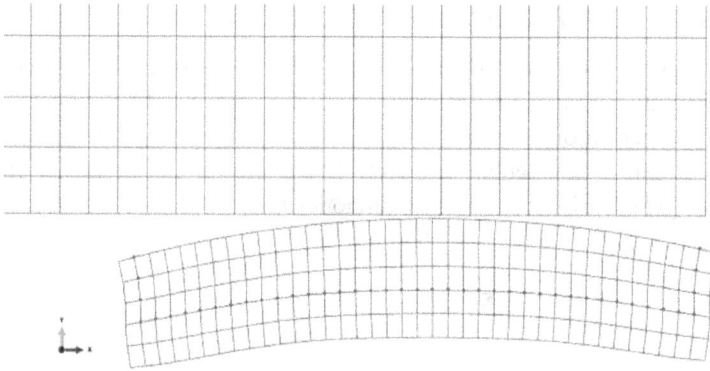

Fig. 7: Lagrangian-type adaptive mesh constraints applied on the node set lagrange_bound_ nodes.

The option USER must be included in the adaptive mesh constraints definition to link the simulation with the user subroutine UMESHMO-TION, which in turn calculates the mesh motion and returns its value to Abaqus.

The boundaries of the adaptive mesh domain with its neighbouring regular Lagrangian mesh may be set by defining Lagrangian-type adaptive mesh constraints on a node set (node set name: lagrange_bound_nodes); this node set must contain the shared nodes between the adaptive mesh domain and the regular mesh as depicted in Fig. 7.

The boundaries between the adaptive mesh domain and the regular mesh may be defined as follows:

```
*ADAPTIVE MESH CONSTRAINT, CONSTRAINT TYPE=LAGRANGIAN
lagrange_bound_nodes
```

Note that the names assigned for the node and element sets are merely random and, although must follow the naming convention of sets in Abaqus, are not preset.

The amount of the volume change due to wear in the adaptive domain is available using the history output variable VOLC, which may be requested as follows:

```
*OUTPUT, HISTORY
*ELEMENT OUTPUT, ELSET=adaptive_elem
VOLC
```

6.1.4. *Mesh motion (ablation equation)*

As previously stated, material ablation will be modelled in the pin according to the Holm–Archard [13, 14] wear equation, whereby the volumetric wear rate is a function of the wear coefficient (K), the contact pressure (p) and the slip rate ($\dot{s}$). To calculate the overall wear from Eq. (7), the rate of nodal ablation ($\dot{h}$) must be summed over all nodes ($1 \ldots n$) within the contact area as follows:

$$\sum_{1}^{n} \dot{h}_i A_i = K \sum_{1}^{n} p_i \dot{s}_i A_i, \tag{9}$$

which, in an FE framework, can be rewritten in terms of an incremental nodal change in height (dh_i) as follows:

$$dh_i = K p_i ds_i, \tag{10}$$

where ds_i is the incremental nodal slip. The total volumetric wear will be the sum of incremental nodal ablation over all nodes ($i = 1 \cdots n$) within the contact area.

The UMESHMOTION subroutine is as follows:

```
1           SUBROUTINE UMESHMOTION(UREF,ULOCAL,NODE,NNDOF,
2        $      LNODETYPE,ALOCAL,NDIM,TIME,DTIME,PNEWDT,
3        $      KSTEP,KINC,KMESHSWEEP,JMATYP,JGVBLOCK,LSMOOTH)
4     C
5           INCLUDE 'ABA_PARAM.INC'
6     C
7           CHARACTER*80 PARTNAME
8           DIMENSION ULOCAL(NDIM)
9           DIMENSION ALOCAL(NDIM,*)
10          DIMENSION JGVBLOCK(*), JMATYP(*)
11          PARAMETER (NELEMMAX=100)
12          DIMENSION JELEMLIST(NELEMMAX), JELEMTYPE(NELEMMAX)
13          PARAMETER (maxNodeNumber=400)
14          DIMENSION ARRAY(600)
15          DIMENSION XCSLIPOLD(maxNodeNumber)
16          DIMENSION XSLIPINC(maxNodeNumber)
17          REAL XCPRESS
18          REAL XCSLIP
19          REAL SURFV
20          COMMON /sequence/
21        $      XTESTINC,XTESTSWEEP
22          DATA XCSLIPOLD /maxNodeNumber*0.0/
23          DATA XSLIPINC /maxNodeNumber*0.0/
```

```
24          DATA nodeNumber /1/
25          DATA XTESTINC /-1/
26   C
27          NELEMS=NELEMMAX
28          JTYP=0
29          CALL GETNODETOELEMCONN(NODE,NELEMS,JELEMLIST,JELEMTYPE,
30       $     JRCD,JGVBLOCK)
31   C
32          IF (KINC.EQ.1.AND.KMESHSWEEP.EQ.0.AND.XTESTINC.EQ.-1) THEN
33            XTESTINC=2
34            XTESTSWEEP=1
35          ELSEIF (KINC.EQ.XTESTINC) THEN
36            XTESTINC=KINC+1
37            XTESTSWEEP=1
38            nodeNumber=1
39          ENDIF
40          IF (KMESHSWEEP.EQ.XTESTSWEEP) THEN
41            XTESTSWEEP=KMESHSWEEP+1
42            nodeNumber=1
43          ENDIF
44   C
45          LOCNUM=0
46          JRCD=0
47          PARTNAME='   '
48   C
49          CALL GETPARTINFO(NODE,JRCD,PARTNAME,LOCNUM,JRCD)
50   C
51          CALL GETVRMAVGATNODE(LOCNUM,JTYP,'CSTRESS',ARRAY,JRCD,
52       $     JELEMLIST,NELEMS,JMATYP,JGVBLOCK)
53          XCPRESS=ARRAY(1)
54   C
55          CALL GETVRMAVGATNODE(LOCNUM,JTYP,'CDISP',ARRAY,JRCD,
56       $     JELEMLIST,NELEMS,JMATYP,JGVBLOCK)
57          XCSLIP=ARRAY(2)
58          XSLIPINC(nodeNumber)=XCSLIP-XCSLIPOLD(nodeNumber)
59          XCSLIPOLD(nodeNumber)=XCSLIP
60   C
61          SURFV=UREF*XCPRESS*XSLIPINC(nodeNumber)
62   C
63   C      Applying wear
64   C
65          DO KDIM=1, NDIM
66            DO JDIM=1, NDIM
67               ULOCAL(KDIM)=ULOCAL(KDIM)-ALOCAL(JDIM,KDIM)*SURFV
68            ENDDO
69          ENDDO
```

```
70  C
71          nodeNumber=nodeNumber+1
72  C
73          RETURN
74          END
```

Declaring variables (lines 7–25)

In comparison to Section 5.2, two additional statements were used in this subroutine, namely, COMMON and DATA. The COMMON statement specifies that certain variables will use the main memory storage and are shared among the various program units. In this example, the common block has the name "sequence" and contains two variables: XTESTINC and XTESTSWEEP. This is a necessary step to enable UMESHMOTION to retrieve data stored in a previous increment or mesh sweep. The DATA statement is simply a method to input data that are known at the time of program writing. In addition to setting a value for the variables node-Number and XTESTINC, it was also used here to initialise the arrays XCSLIPOLD and XSLIPINC.

Obtaining contact pressure and slip (lines 51–59)

The computation of incremental nodal ablation requires pre-knowledge of the values of both the nodal contact pressure (p_i) and the nodal incremental slip (ds_i). The former may be simply obtained by calling the utility subroutine GETVRMAVGATNODE with the *output variable identifier* CSTRESS. The incremental slip, on the other hand, cannot be readily obtained from Abaqus without an extra programming procedure. The values of the relative tangential motion (i.e. slip) are stored into the accumulated outputs CSLIP1 and CSLIP2 in 3D models and in CSLIP1 in 2D models. These values are calculated by summing the scalar product of the incremental relative nodal displacement vector and a local tangent direction. The incremental relative nodal displacement vector is a measure of the motion of a slave node relative to the motion of the master surface during closed contact. Accordingly, for this example, the incremental slip values must be obtained from the accumulated variable CSLIP1, taking into account the slip direction, and stored into a separate variable. The accumulated slip may be obtained by calling the utility subroutine GETVRMAVGATNODE with the *output variable identifier* CDISP. By referring to the section "Abaqus/Standard output variable identifiers/Surface variables" of the Abaqus Output Guide [10] it shows that the variable CDISP (available as output for .fil file format) returns the following individual attributes:

- COPEN (separation of the surfaces in the direction normal to the master surface),
- CSLIP1 (accumulated relative tangential displacement of the surfaces in the local 1-direction on the master surface), and
- CSLIP2 (accumulated relative tangential displacement of the surfaces in the local 2-direction on the master surface for 3D).

In order to obtain incremental slip values from the accumulated value CSLIP1, the latter was initially retrieved from the second element of the array variable ARRAY, i.e. ARRAY(2), and stored into a pre-defined variable (XCSLIP) as the current accumulated slip. Subsequently, the absolute value resulting from the subtraction of the accumulated slip stored in the previous increment or mesh sweep (XCSLIPOLD) from its current value (XCSLIP) was calculated and stored into XSLIPINC. The information regarding the sequence of increments and mesh sweeps was obtained by programming *lines 32–43*.

It should be noted that incremental slip values should be obtained on nodal basis. Hence, by adopting this procedure, nodal slip data were stored into arrays following a virtual nodal arrangement, defined by the variable nodeNumber (*lines 32–43* and *line 71*), thus, yielding the arrays XSLIP-INC(nodeNumber) and XCSLIPOLD(nodeNumber) to store the nodal incremental slip and the nodal accumulated slip, respectively.

In 3D modelling, the slip magnitude may be obtained from its two components CSLIP1 and CSLIP2; the value is computed as follows: $|Slip| = \sqrt{CSLIP_1^2 + CSLIP_2^2}$. The same applies to computations dependent on the magnitude of the surface shear traction, which may be calculated from its two components CSHEAR1 and CSHEAR2.

Calculating wear and ablating nodes (lines 32–40)

The computation of wear and ablation of nodes was executed using the same procedure explained in Section 5.2.

6.1.5. *Reading and understanding the results*

As previously mentioned in Section 6.1.3, Abaqus/Standard provides the variable VOLC as a history output. It gives the amount of volume change due to wear in a defined element set. Requesting VOLC for the adaptive mesh domain, one would obtain the wear volume calculated throughout the simulation as a function of simulation time. In case of material ablation, the change in volume will be negative, thus, indicating material loss due to wear.

By running the simulation detailed in this example, the change in volume VOLC in the element set `adaptive_elem` may be plotted against the simulation time as shown in Fig. 8. The results indicate a linear relationship between the wear volume and time, and since the sliding distance is proportional to time, this yields a linear relationship of wear volume with the sliding distance as well.

This proportionality between the wear volume and sliding distance, which is common in systems modelled with the Holm–Archard wear equation, may be manipulated by numerical extrapolation of wear data for longer sliding distances, thus, eliminating the need to run long computationally expensive simulations. Although this method will provide data for the wear volume, information pertaining to geometric alterations and the consequent changes in the stress state will not be available.

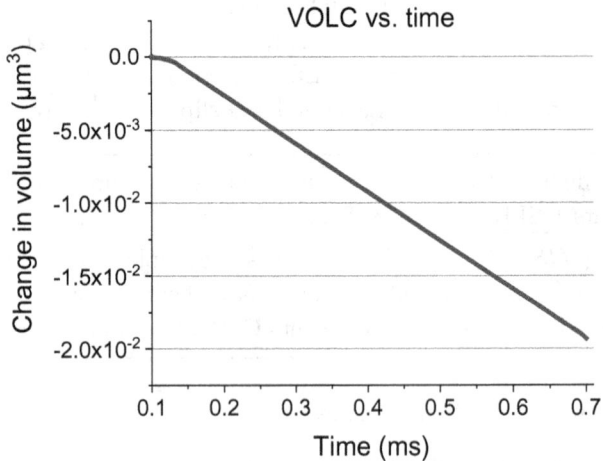

Fig. 8:   Change in volume VOLC in the element set `adaptive_elem`.

7. Practical Considerations

7.1. *Defining the mesh motion velocity (wear coefficient)*

As previously mentioned, the mesh motion velocity (or the wear coefficient) may be set by defining the mesh motion constraint of type velocity as follows:

```
*ADAPTIVE MESH CONSTRAINT, CONSTRAINT TYPE=SPATIAL,
TYPE=VELOCITY, USER
node set,,, magnitude of the mesh motion velocity
```

The reliable method of obtaining a value for the mesh motion velocity is by running tribological tests under the same tribological conditions (i.e. contact configuration, temperature, lubrication, etc.). Defining a single wear coefficient lumps all intertwined wear mechanisms that might be otherwise very difficult to discern into one single quantity. Here, we take an example of the wear of silicon nitride rollers in rolling-sliding contact against hardened bearing steel in a double-roller configuration as treated in [15]. To determine the wear coefficient (K, as established in Eq. (8)), it is best to plot the wear volume measured on the sample against the force multiplied by the sliding distance; see Fig. 9, plot of the wear volume.

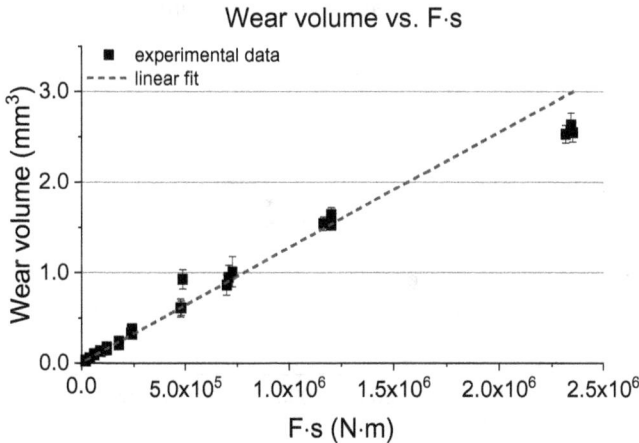

Fig. 9: Plot of the wear volume V_w against the applied normal force multiplied by the sliding distance $F \cdot s$.

A linear regression of the data will result in a line whose slope is the wear coefficient. The data yield a slope for the V_w *vs.* $F \cdot s$ diagram of $K = 1.27 \times 10^{-6}$ mm^3/N$\cdot$ m, which should be converted into the equivalent units to match those adopted in the FE model; this results in a wear coefficient of $K = 1.27 \times 10^{-9}$ mm^3/N$\cdot$ mm.

If plotting a V_w *vs.* $F \cdot s$ diagram is not possible due to the lack of experimental data, an adapted wear coefficient may be obtained by matching the wear volume obtained from the tribological test with the value of VOLC obtained from the simulation. This is an iterative process and, thus, will require running the simulation several times until reaching wear rates that are close enough to those obtained experimentally. Eventually, the wear volume data as function of time from both the tribological tests and the wear simulations should give comparable result as shown in the V_w *vs.* t diagram in Fig. 10. It is important to note that the error margin resulting from adopting this method may be considerable if the test time and the simulation time significantly differ; in this case, interpolation or extrapolation of experimental data becomes inevitable to match the results.

On several occasions, it has been recognised that simulating wear by relying on one single wear coefficient may lead to inaccurate results, such as significant disparity between the worn geometry obtained experimentally and that resulting from the simulation. In such cases, it might be more

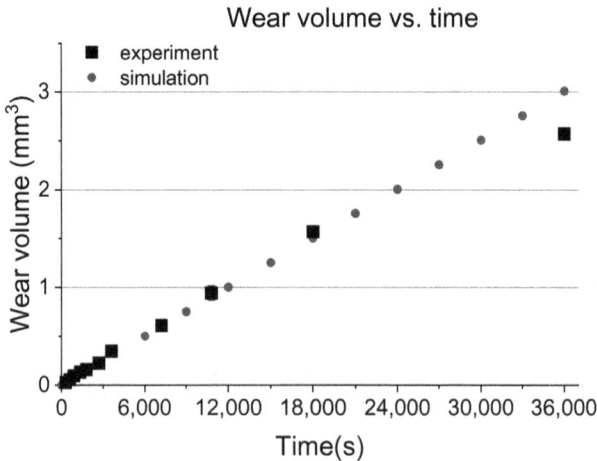

Fig. 10: Plot of the wear volume V_w against time t.

adequate to adopt a wear equation that incorporates multiple wear coefficients, each of which governs the influence of a solution specific quantity, e.g. temperature, stress, etc., on the rate of ablation. Here it should be reminded that only one coefficient may be passed from Abaqus to the subroutine. All other coefficients must be defined from within the subroutine itself.

7.2. *Accelerating wear*

In order to reduce the computational cost of a wear simulation, the user may be tempted to accelerate wear by applying a scaling factor, either to the mesh motion velocity (i.e. the wear coefficient) or the local wear equation. While this may prove helpful in certain cases, a sensitivity analysis must be conducted to verify the quality of the results. Too high a material recession velocity may result in several undesired issues such as severely intermittent contacts, caused by losing local contacts at nodes, thus, causing conversion difficulties or even resulting in inhomogeneous material recession leading eventually to severe mesh distortions and unusable results. Hence, a scaling factor must only be used after careful consideration of its effects on the quality of the deformed mesh.

8. Concluding Remarks

The FEM allows incorporating several phenomena into one simulation model. The main purpose of modelling wear in an FE framework is to obtain information pertaining to the evolution of the contact area, contact stresses and interface properties as a function of the modified geometry.

Abaqus/Standard provides the possibility to model wear using the user subroutine UMESHMOTION, which is based on the ALE adaptive meshing. Defining the mesh motion velocity (i.e. wear coefficient) relies on running tribological tests, from which wear rates are obtained. The accuracy of the wear coefficient is dependent on the level by which the tribological test mimics the system under consideration. An iterative process in comparing the wear rates obtained from the simulation with those obtained from tribological tests would yield an acceptable representation of the system.

Finally, if modelling the system using one single wear coefficient is deemed inadequate, other coefficients may be programmed from within the subroutine.

References

[1] Dassault Systèmes, SIMULIA User Assistance 2018: Abaqus Interactions Guide, Dassault Systèmes Simulia Corp., 2017.

[2] J. Larsen-Basse, Basic Theory of Solid Friction, in G. Totten (ed.), *ASM Handbook, Volume 18: Friction, Lubrication and Wear Technology*. ASM International, 1992.

[3] K. Johnson, Influence of interfacial friction, in *Contact Mechanics*. Cambridge, Cambridge University Press, 1985, pp. 119–124.

[4] J. Halme, P. Andersson, Rolling contact fatigue and wear fundamentals for rolling bearing diagnostics — state of the art. *Proceedings of the Institution of Mechanical Engineers, Part J Eng Tribol. 2010*, **224**(4), 377–393. doi:10.1243/13506501JET656

[5] Dassault Systèmes, *SIMULIA User Assistance 2018: Abaqus Analysis Guide*. Dassault Systèmes Simulia Corp., 2018.

[6] J. Oden and J. Martins, Models and computational methods for dynamic friction phenomena, *Comput Method Appl Mech Eng.* **52**(1–3), 527–634 (1985).

[7] S. Kato and T. Matsubayashi, On the dynamic behaviour of machine tool slideway: 1st report, characteristics of static friction in stick slip motion, *Bull JSME.* **13**(55), 170–179 (1970).

[8] S. Kato, N. Sato, and T. Matsubayashi, Some considerations on characteristics of static friction of machine tool slideway, *ASME J Lubr Technol.* **94**(3), 234–247 (1972).

[9] Dassault Systèmes, *SIMULIA User Assistance 2018: Abaqus User Subroutines Guide*. Dassault Systèmes Simulia Corp., 2017.

[10] Dassault Systèmes, *SIMULIA User Assistance 2018: Abaqus Output Guide*. Dassault Systèmes Simulia Corp., 2017.

[11] I. Khader, A. Renz, and A. Kailer, A wear model for silicon nitride in dry sliding contact against a nickel-base alloy, *Wear.* **376–377**, 352–362 (2017).

[12] B. Zhao, I. Khader, R. Raga, U. Degenhardt, and A. Kailer, Tribological behaviour of three silicon nitride ceramics in dry sliding contact against Inconel 718 over a wide range of velocities, *Wear.* **448–449**, 203–206 (2020).

[13] R. Holm, *Electric Contacts*. Stockholm: Almquist and Wiksells Akademiska Handböcker, 1946.

[14] J. Archard, Contact and rubbing of flat surfaces, *J Appl Phys.* **24**, 981–988 (1953).

[15] I. Khader, D. Kürten, and A. Kailer, A study on the wear of silicon nitride in rolling-sliding contact, *Wear.* **296**(1–2), 630–637 (2012).

Chapter 8

Wear Modelling in Fibre-Reinforced Composite Materials

J. M. Juliá*, L. Rodríguez-Tembleque*,‡ and M. H. Ferri Aliabadi†

*Escuela Técnica Superior de Ingeniería, Universidad de Sevilla,
Camino de los Descubrimientos s/n, Sevilla 41092, Spain
†Department of Aeronautics, Faculty of Engineering,
Imperial College of London,
South Kensington Campus, London SW7 2AZ, UK
‡luisroteso@us.es

This chapter presents a computational framework for fretting wear simulation in aligned fibre-reinforced composite materials. Tribology of fibre-reinforced composites is a difficult task due to their friction and wear mechanisms being more complex than in metal. Among several factors, friction and wear constitutive laws require considering not only the normal contact pressure and the sliding velocity, but also the temperature and the micromechanical aspects such as the fibre volume fraction or the fibre orientation relative to the sliding direction. As a first approach to this issue, this work presents a boundary element-based 3D formulation to simulate fretting wear in fibre-reinforced composites including the influence of micromechanics. Consequently, phenomenological friction and wear laws are developed to take into account those micromechanical aspects. These tribological laws have been incorporated into an augmented Lagrangian resolution scheme and applied to compute and study wear in a carbon fibre-reinforced composite.

1. Introduction

Fibre-reinforced composite materials are wildly used in many structural and mechanical systems in Aerospace, Automobile, Biomedical[1], Building and Civil [2] engineering. In many of these applications, these materials are subjected to contact and interface loads (e.g. mechanical joints between fibre-reinforced composite profiles and stainless steel connections or contact pin-loaded holes in fibre-reinforced composite plates), especially in those engineering applications in which friction and wear are critical issues [3], i.e. in tribo-components (e.g. roller bearing cages). So nowadays engineers require more sophisticated numerical tools that make it possible to study these materials and design these components under contact and wear conditions.

Friction and wear behaviour of fibre-reinforced plastics (FRP) has been studied in depth since the end of the second half of the 20th century in the following works [4–15]. These experimental contributions have studied the significant influence of fibre orientation on the wear and frictional behaviour of FRP composites. Those works showed that the friction coefficient depends on several factors including the combination of materials, the surface roughness or the fibre orientation (i.e. the largest coefficient of friction is obtained when the sliding is normal to the fibre orientation, while the lowest one is obtained when the fibre orientation is parallel to the direction of sliding) (see Fig. 1(a)). Considering a sliding direction on a plane parallel to the direction of fibres, Ohmae *et al.* [4] observed that the friction coefficient sliding in parallel direction was smaller than in the transverse direction ($\mu_L \leq \mu_T \leq \mu_N$). So experiments reveal that contact and wear constitutive laws have to consider not only the micromechanics of the anisotropic bulk but also the fibres orientation (φ) and the sliding direction (θ) (see Fig. 1(b)).

Several semi-analytical works [16–24] have dealt with the problem of contact and interaction modelling of FRP. However, due to their intrinsic mathematical complexity, analytical solutions incorporate several restrictive assumptions (e.g. rigid indenter, half-plane space or the sliding direction presented in Fig. 1(a)).

Numerical solutions based on the finite element method (FEM) [26, 27] started to study some contact problems between composites, but isotropic friction laws were assumed. The indentation problem of fibre-reinforced polymer was initially studied by Varadi *et al.* [28]. Subsequently, an FEM formulation involving macro- and micro-contact analyses was presented in Varadi *et al.* [29], and more recently, the fibre–matrix debonding process was studied in Goda *et al.* [30, 31]. As it can be observed in these works, a very fine mesh must be considered to approximate the contact problem between these composite domains.

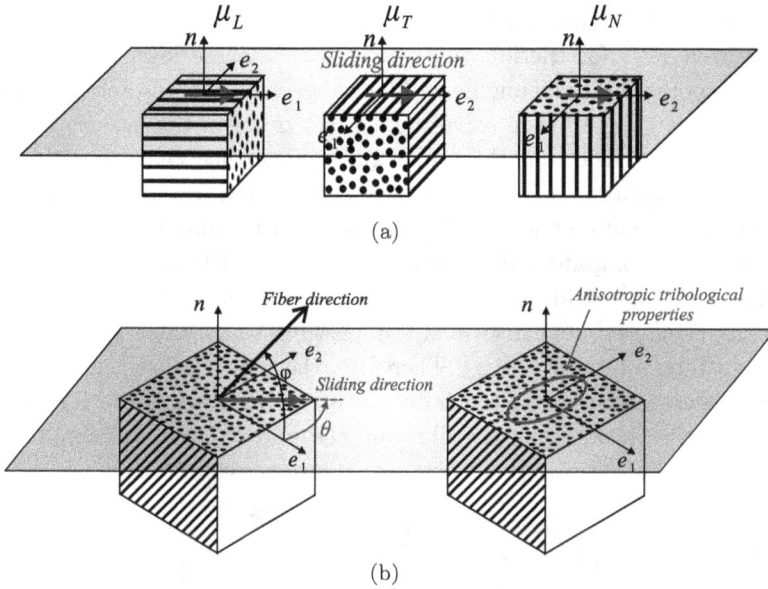

Fig. 1: (a) Fibre-reinforced material indicating the sliding directions (i.e. longitudinal, transverse and normal) and their corresponding friction coefficients: μ_L, μ_T and μ_N. (b) Unidirectional FRP with arbitrary fibre and sliding directions. (These figures were reprinted from Rodríguez-Tembleque and Aliabadi [25] with permission from Elsevier.)

The boundary element method (BEM) [32] has been a well-recognised, generally very accurate and efficient numerical tool for studying contact and interface problems in FRP materials. Many works have applied the BEM to study these problems in FRP, because the BEM has proved to obtain a very good accuracy with less number of elements than finite element methodologies. The fibre–matrix debonding problem was considered by Varna et al. [33], Graciani et al. [34, 35] and Tavara et al. [36]. Optimisation/identification analysis of inclusions (i.e. fibres) in frictionless unilateral contact with the matrix was dealt with by Mallardo et al. [37] and fibre-reinforced composites under frictional indentation problems were studied in Rodríguez-Tembleque et al. [38, 39]. However, all these numerical works solve different contact problems assuming frictionless contact, or constant friction (and wear) coefficients, which are, for example, independent of fibre orientation and sliding direction.

This chapter presents a 3D boundary element formulation for fretting wear in fibre-reinforced composites. This numerical framework was presented in Rodríguez-Tembleque *et al.* [25, 40] and proposes new contact constitutive laws for friction and wear in FRP. These laws are presented and incorporated into an augmented Lagrangian contact resolution scheme, which makes it possible to solve the contact problem taking into account both the mechanical and the tribological anisotropic characteristics (i.e. anisotropic bulk properties and anisotropic wear and frictional conditions). The proposed contact and wear laws, as well as the numerical methodology, are applied to compute and study wear in carbon FRP films. Furthermore, the formulation considers a micromechanical model for FRP presented in Hopkins *et al.* [41] that also makes it possible to consider micromechanics (e.g. fibre volume fraction). Therefore, the numerical experiments will present several studies on the influence of: fibres orientation, the sliding direction, the thickness of the film and the fibre volume fraction, on the normal and tangential contact forces or the wear evolution.

2. Governing Equations

Let us consider two homogeneous linearly anisotropic solids in contact (see Fig. 2). They occupy the regions $\Omega^l \subset \mathbb{R}^3$ ($l = 1, 2$) with boundary $\partial\Omega^l$ in a Cartesian coordinate system (x_i) ($i = 1, 2, 3$). The boundary of each solid $\partial\Omega^l$ is divided into three disjoint parts: $\partial\Omega^l_t$ with prescribed tractions $\tilde{t}^l_i$, $\partial\Omega^l_u$ with imposed displacements $\tilde{u}^l_i$ and $\partial\Omega^l_c$ representing the potential contact surfaces, which have outward unit normal vectors n^l_i. Under small displacement assumption, these boundaries are almost coincident (i.e. $\partial\Omega^1_c \simeq \partial\Omega^2_c$) so we can define a common contact surface $\partial\Omega_c$ with a normal vector $n_{c,i} \simeq n^1_i \simeq n^2_i$.

On the domains Ω^l, the mechanical equilibrium equations in the absence of body forces are

$$
\begin{aligned}
\sigma_{ij,j} &= 0 && \text{in} \quad \Omega^1 \cup \Omega^2, \\
\sigma_{ij} n_j &= \tilde{t}_i && \text{on} \quad \partial\Omega^1_t \cup \partial\Omega^2_t, \\
\sigma^1_{ij} n_{c,j} &= -\sigma^2_{ij} n_{c,j} = p_i && \text{on} \quad \partial\Omega_c,
\end{aligned}
\tag{1}
$$

σ_{ij} being the components of Cauchy stress tensor, n_i the unit normal on $\partial\Omega^1_t \cup \partial\Omega^2_t$, p_i is the contact traction and σ^1_{ij} and σ^2_{ij} are restrictions of σ_{ij} to a particular domain Ω^l_t. In this chapter, we consider the small strain hypothesis so the infinitesimal strain tensor ε_{ij} is defined as: $\varepsilon_{ij} = (u_{i,j} + u_{j,i})/2$ in $\Omega^1 \cup \Omega^2$, where u_i are the components of the elastic displacement field in $\Omega^1 \cup \Omega^2$. The stress and strain tensors are coupled through the linear

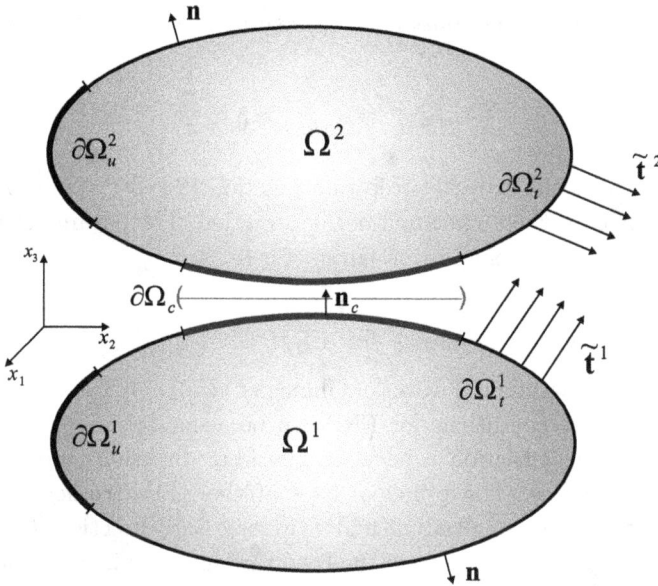

Fig. 2: The physical setting: two solids in contact.

constitutive law

$$\sigma_{ij} = C_{ijkl}\varepsilon_{kl}, \tag{2}$$

where C_{ijkl} denotes the components of the elastic stiffness tensor, which satisfies the following symmetries: $C_{ijkl} = C_{jikl} = C_{ijlk} = C_{klij}$. Moreover, the elastic stiffness tensor is positive definite.

2.1. *Unilateral contact law*

The unilateral contact law involves Signorini's contact conditions in $\partial\Omega_c$

$$g_n \geq 0, \quad p_n \leq 0, \quad g_n \, p_n = 0, \tag{3}$$

where $p_n = \mathbf{p} \cdot \mathbf{n}_c$ is the normal contact pressure and g_n is normal contact gap. The contact gap is defined as

$$g_n = g_o + \omega + u_n, \tag{4}$$

ω being the wear gap (i.e. *wear depth*) and u_n, the relative normal displacement: $u_n = (\mathbf{u}^2 - \mathbf{u}^1) \cdot \mathbf{n}_c$.

The normal contact constraints (3) can be formulated in a more compact form, as follows:

$$p_n - \mathbb{P}_{\mathbb{R}_-}(p_n^*) = 0, \tag{5}$$

where $\mathbb{P}_{\mathbb{R}_-}(\bullet)$ is the normal projection function $(\mathbb{P}_{\mathbb{R}_-}(\bullet) = \min(0, \bullet))$ and $p_n^* = p_n + r_n g_n$ is the augmented normal traction. The parameter r_n is the normal dimensional penalisation parameter $(r_n \in \mathbb{R}^+)$.

2.2. *Frictional contact laws for FRP*

According to experimental works of Ohmae *et al.* [4] and Tsukizoe *et al.* [6], frictional contact conditions for FRP can be accurately approximated by a convex elliptical friction cone when the fibre directions are parallel to the contact surface. The principal axes of the ellipse coincide with the orthotropic axes, i.e. longitudinal and transverse fibre directions (see Fig. 3).

Theoretical investigations in the literature have studied and developed different friction models, which define the admissible region for contact tractions (Friction Cone) and the sliding rules. A general theory of friction was presented by Curnier in [42]. In this way, Michalowski and Mróz [43], Mróz and Stupkiewicz [44] and Mróz *et al.* [45] considered an orthotropic model with associated and non-associated sliding rules. This work considers an associated sliding rule formulated in the same way as Feng *et al.* [46, 47].

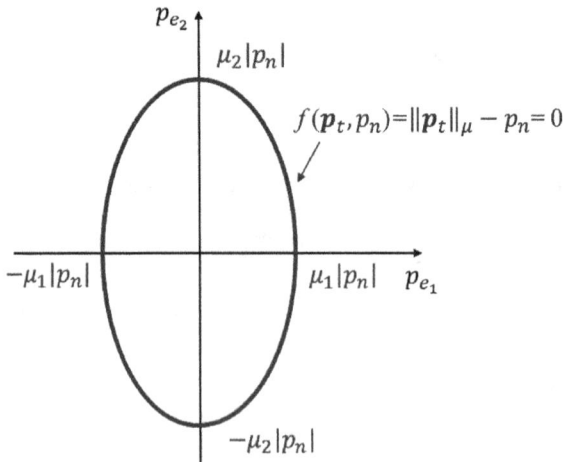

Fig. 3:   Elliptic friction law.

The generic form of the friction surface presented in Fig. 3 is

$$f(\mathbf{p}_t, p_n) = ||\mathbf{p}_t||_\mu - |p_n| = 0, \tag{6}$$

where $\mathbf{p}_t = \mathbf{p} - p_n\mathbf{n}_c$ and $|| \bullet ||_\mu$ denotes the elliptic norm

$$||\mathbf{p}_t||_\mu = \sqrt{(p_{e_1}/\mu_1)^2 + (p_{e_2}/\mu_2)^2}, \tag{7}$$

μ_1 and μ_2 being the principal friction coefficients in the directions $\{e_1, e_2\}$.

The Coulomb friction restriction ($||\mathbf{p}_t||_\mu \leq |p_n|$) can be summarised as follows:

$$||\mathbf{p}_t||_\mu < |p_n| \Rightarrow \dot{\mathbf{g}}_t = \mathbf{0} \quad \text{on } \partial\Omega_c, \tag{8}$$

$$||\mathbf{p}_t||_\mu = |p_n| \Rightarrow \mathbf{p}_t = -|p_n|\mathbb{M}^2\dot{\mathbf{g}}_t/||\dot{\mathbf{g}}_t||_\mu^* \quad \text{on } \partial\Omega_c. \tag{9}$$

In the expressions above, the tangential slip velocity $\dot{\mathbf{g}}_t$ can be expressed at time τ_k as follows: $\dot{\mathbf{g}}_t \approx \Delta\mathbf{g}_t/\Delta\tau$ (see [48]), where $\Delta\mathbf{g}_t = \mathbf{g}_t(\tau_k) - \mathbf{g}_t(\tau_{k-1})$, $\Delta\tau = \tau_k - \tau_{k-1}$ and $\mathbf{g}_t = (\mathbf{u}^2 - \mathbf{u}^1) - u_n\mathbf{n}_c$, defining the initial tangential slip as zero.

In Eq. (9), the value for the tangential contact traction was presented in [48] assuming an associated sliding rule, the norm $|| \bullet ||_\mu^*$ is dual of $|| \bullet ||_\mu$, i.e.

$$||\dot{\mathbf{g}}_t||_\mu^* = \sqrt{(\mu_1\dot{g}_{e_1})^2 + (\mu_2\dot{g}_{e_2})^2}, \tag{10}$$

and $\mathbb{M}$ is a diagonal matrix:

$$\mathbb{M} = \begin{bmatrix} \mu_1 & 0 \\ 0 & \mu_2 \end{bmatrix}. \tag{11}$$

The fibres can be oriented with any angle ($0 \leq \varphi \leq \pi/2$) relative to direction e_1, as defined in Fig. 1 (b). So the principal friction coefficients in the directions $\{e_1, e_2\}$ have to be defined as

$$\mu_1 = \mu_L + (\mu_N - \mu_L)\,\hat{\varphi}, \tag{12}$$

$$\mu_2 = \mu_T + (\mu_N - \mu_T)\,\hat{\varphi}, \tag{13}$$

where the parameter

$$\hat{\varphi} = 2\varphi/\pi \tag{14}$$

is the non-dimensional fibre orientation angle ($0 \leq \hat{\varphi} \leq 1$), and $\{\mu_L, \mu_T, \mu_N\}$ are the friction coefficients in longitudinal, transverse and normal directions, respectively.

The values for μ_L, μ_T and μ_N should be obtained from experimental works. In this case, we have considered the values from Ohmae *et al.* [4] and Tsukizoe and Ohmae [6]. So the anisotropic friction surface Eq. (6) is also a function of the fibre orientation angle ($\hat{\varphi}$):

$$f(\mathbf{p}_t, p_n, \hat{\varphi}) = ||\mathbf{p}_t||_{\mu(\hat{\varphi})} - |p_n| = 0. \tag{15}$$

Consequently, this friction law adopts the form presented in Fig. 4.

We can see in Fig. 4 how an orthotropic friction law is obtained when the fibres are parallel to the sliding plane ($\hat{\varphi} = 0$), and an isotropic friction law is obtained when the fibres are normal to the sliding plane ($\hat{\varphi} = 1$). This behaviour agrees with the experiments presented by Ohmae *et al.* [4] and Sung *et al.* [5]. So, when fibres are parallel to the sliding plane (i.e. $\varphi = 0$), the principal friction coefficients (μ_1 and μ_2) values are: μ_N and μ_T. However, when $\varphi \neq 0$, their values are approximated by Eq. (12) and Eq. (13). In case $\varphi = \pi/2$, the isotropic friction law is recovered, i.e. $\mu_1 = \mu_2 = \mu_N$.

The frictional contact constraints (8)–(9) can be also formulated using contact operators as follows:

$$\mathbf{p}_t - \mathbb{P}_{\mathbb{E}_\rho}(\mathbf{p}_t^*) = 0, \tag{16}$$

where $\mathbf{p}_t^* = \mathbf{p}_t - r_t \mathbb{M}^2 \dot{\mathbf{g}}_t$ ($r_t \in \mathbb{R}^+$) is the augmented tangential traction and $\mathbb{P}_{\mathbb{E}_\rho}(\bullet) : \mathbb{R}^2 \longrightarrow \mathbb{R}^2$ is the tangential projection function defined in [48] as

$$\mathbb{P}_{\mathbb{E}_\rho}(\mathbf{p}_t^*) = \begin{cases} \mathbf{p}_t^* & \text{if} \quad ||\mathbf{p}_t^*||_\mu < \rho, \\ \rho\, \mathbf{p}_t^*/||\mathbf{p}_t^*||_\mu & \text{if} \quad ||\mathbf{p}_t^*||_\mu \geq \rho, \end{cases} \tag{17}$$

with $\rho = |\mathbb{P}_{\mathbb{R}_-}(p_n^*)|$.

2.3. *Wear law for FRP*

A quasi-steady-state wear approximation similar to the computational works [49–57] is considered. So wear law can be expressed in the following wear rate form:

$$\dot{\omega} = i_w\, |p_n|\, ||\dot{\mathbf{g}}_t||, \tag{18}$$

where i_w is the dimensional wear coefficient.

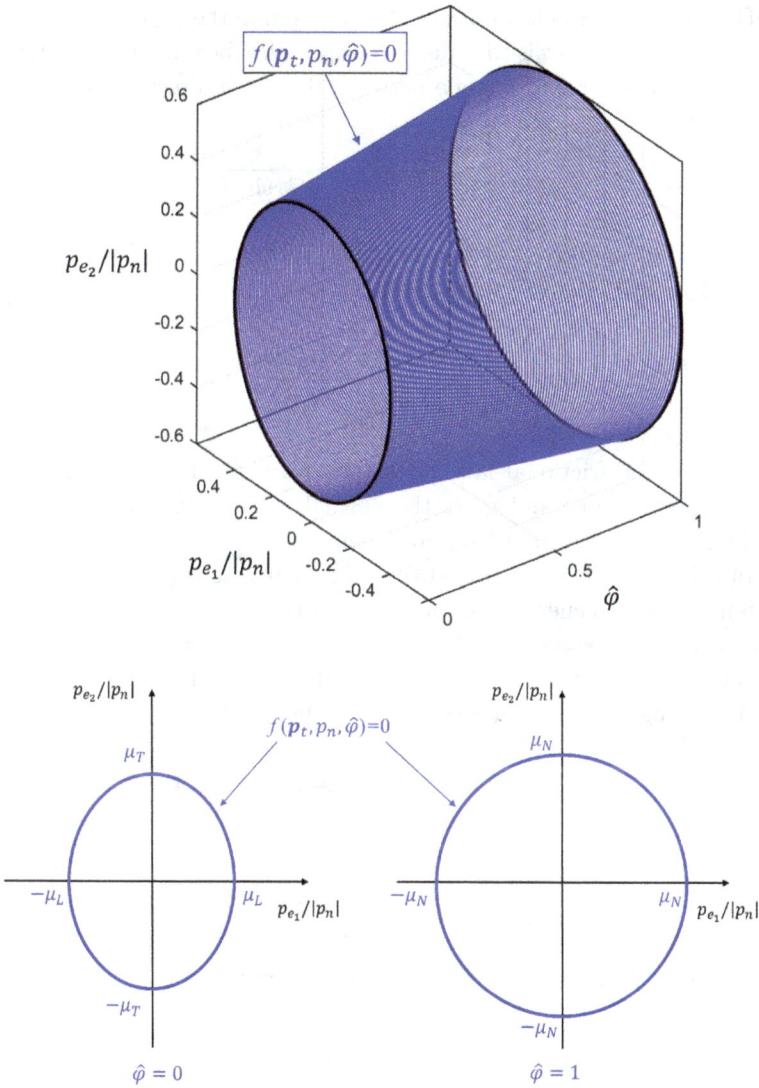

Fig. 4: The friction surface representation $f(\mathbf{p}_t, p_n, \hat{\varphi})/|p_n| = 0$ for $\mu_L = 0.4$, $\mu_T = 0.5$, and $\mu_N = 0.6$, whose axes were defined as $\{p_{e_1}/|p_n|, p_{e_2}/|p_n|, \hat{\varphi}\}$. Frictional response of the FRP surface depends on the sliding direction $\{p_{e_1}/|p_n|, p_{e_2}/|p_n|\}$ and the fibre orientation $\hat{\varphi}$. So, an elliptic friction law, whose principal friction coefficients are $\mu_1 = \mu_L$ and $\mu_2 = \mu_T$, is obtained when the fibres are parallel to the sliding plane ($\hat{\varphi} = 0$). An isotropic friction law (i.e. $\mu_1 = \mu_2 = \mu_N$) is obtained when the fibres are normal to the sliding plane ($\hat{\varphi} = 1$).

Due to the wear behaviour of FRP observed in the experimental works, an orthotropic wear law is considered. So i_w should be function of the sliding direction parameter θ (i.e. angle between the given direction ($\mathbf{e}_1$) and the sliding direction)

$$i_w(\theta) = \sqrt{(i_1 \cos\theta)^2 + (i_2 \sin\theta)^2}. \tag{19}$$

In the expression above, $\cos\theta = \dot{g}_{e_1}/||\dot{\mathbf{g}}_t||$, $\sin\theta = \dot{g}_{e_2}/||\dot{\mathbf{g}}_t||$, and i_1 and i_2 are the principal intensity wear coefficients:

$$i_1 = i_L + (i_N - i_L)\hat{\varphi}, \tag{20}$$

$$i_2 = i_T + (i_N - i_T)\hat{\varphi}. \tag{21}$$

Similarly to the frictional law, i_L is the longitudinal wear coefficient, i_T is the transverse one and i_N is the normal wear coefficient. Their values should be considered from the experiments.

In this work, it is postulated that the wear rate is proportional to the friction dissipation energy, friction and wear coefficients are related through the constant c as $i_L = c\,\mu_L|p_n|$, $i_T = c\,\mu_T|p_n|$ and $i_N = c\,\mu_N|p_n|$. However, they can also be obtained from experimental works like [4, 6].

According to [52], wear coefficient can be written as

$$i_w = ||\dot{\mathbf{g}}_t||_i/||\dot{\mathbf{g}}_t||, \tag{22}$$

where

$$||\dot{\mathbf{g}}_t||_i = \sqrt{(i_1\dot{g}_{e_1})^2 + (i_2\dot{g}_{e_2})^2}. \tag{23}$$

So the orthotropic wear law Eq. (18) can be rewritten as

$$\dot{\omega} = |p_n|\,||\dot{\mathbf{g}}_t||_i. \tag{24}$$

Finally, for quasi-static contact problems, derivatives can be expressed in an incremental form: $\dot{\omega} \simeq \Delta\omega = \omega^{(k)} - \omega^{(k-1)}$ and $\dot{\mathbf{g}}_t \simeq \Delta\mathbf{g}_t = \mathbf{g}_t^{(k)} - \mathbf{g}_t^{(k-1)}$, therefore, wear depth at the instant k can be computed as

$$\omega^{(k)} = \omega^{(k-1)} + |p_n^{(k)}|\,||\,\mathbf{g}_t^{(k)} - \mathbf{g}_t^{(k-1)}||_i. \tag{25}$$

3. Numerical approximation

3.1. *Boundary element equations*

The well-known elastic boundary integral equation for displacements at the source point $\mathbf{x}' \in \partial\Omega$ [32], in the absence of body forces, can be written as

$$c_{jk}(\mathbf{x}')u_j(\mathbf{x}') + \fint_{\partial\Omega} \check{T}_{jk}(\mathbf{x}',\mathbf{x})u_j(\mathbf{x})dS(\mathbf{x}) = \int_{\partial\Omega} \check{U}_{jk}(\mathbf{x}',\mathbf{x})t_j(\mathbf{x})dS(\mathbf{x}), \quad (26)$$

where $\mathbf{x}$ is the field point; u_j and t_j are the displacement and the tractions, respectively; $\check{U}_{jk}$ and $\check{T}_{jk}$ are the fundamental solutions. c_{jk} is a constant that depends on the geometry of the boundary at $\mathbf{x}'$ and is equal to $1/2$ for a smooth boundary $\partial\Omega$. The symbol $\fint$ denotes a Cauchy principal value integral.

In this work, the scheme for the evaluation of the fundamental solution for an anisotropic media is the one recently presented in [58]. However, different fundamental solution schemes and implementations can be found in the literature [59–64].

Green's function for anisotropic media can be expressed as a singular term by a modulation function $\check{\mathbf{H}}$ as

$$\check{U}_{jk}(r\hat{\mathbf{e}}) = \frac{1}{4\pi r}\check{H}_{jk}(\hat{\mathbf{e}}), \quad (27)$$

where $r = \|\mathbf{x} - \mathbf{x}'\|$ and $\hat{\mathbf{e}} = (\mathbf{x} - \mathbf{x}')/r$, $\|\cdot\|$ being the *Euclidic norm*. $\check{H}_{jk}(\hat{\mathbf{e}})$ is one of the three Barnett–Lothe tensors which is symmetric and positive-definite. The tensor $\check{H}_{jk}(\hat{\mathbf{e}})$ can be evaluated as

$$\check{H}_{jk}(\hat{\mathbf{e}}) = \frac{1}{\pi}\int_{-\infty}^{+\infty} \Gamma_{jk}^{-1}(p)dp, \quad (28)$$

with $\Gamma_{jk}(p) = Q_{jk} + (R_{jk} + R_{kj})p + T_{jk}p^2$, expressed in terms of the parameter p, and

$$Q_{jk} = C_{ijkl}\hat{n}_i\hat{n}_l \quad R_{jk} = C_{ijkl}\hat{n}_i\hat{m}_l \quad T_{jk} = C_{ijkl}\hat{m}_i\hat{m}_l, \quad (29)$$

where $\hat{n}_i$ and $\hat{m}_i$ are the components of any two mutually orthogonal unit vectors such that $\{\hat{\mathbf{n}}, \hat{\mathbf{m}}, \hat{\mathbf{e}}\}$ is a right-handed triad. Repeated indices imply sum.

The components of the traction fundamental solution follow easily from the derivative of Green's function as

$$\check{T}_{ik} = C_{ijlm} n_j \frac{\partial \check{U}_{lk}}{\partial x_m}, \tag{30}$$

where n_j are the components of the external unit normal vector to the boundary $\partial\Omega$ at point $\mathbf{x}$. The derivative of Green's function may be expressed in a similar way to Eq. (27), as a singular term by a modulation function which only depends on $\hat{\mathbf{e}}$ as

$$\frac{\partial \check{U}_{ij}(r\hat{\mathbf{e}})}{\partial x_q} = \frac{1}{4\pi r^2} \frac{\partial \tilde{U}_{ij}(\hat{\mathbf{e}})}{\partial x_q} \tag{31}$$

where, according to Lee's approach [65], the components of the modulation function are given by

$$\frac{\partial \tilde{U}_{ij}(\hat{\mathbf{e}})}{\partial x_l} = -\hat{e}_l \check{H}_{ij} + \frac{C_{pqrs}}{\pi}(M_{lqiprj}\hat{e}_s + M_{sliprj}\hat{e}_q). \tag{32}$$

The M_{sliprj} are integrals (32) that have the following representation in terms of the parameter p [65]:

$$M_{ijklmn} = \frac{1}{|\mathbf{T}|^2} \int_{-\infty}^{+\infty} \frac{\Phi_{ijklmn}(p)}{(p-p_1)^2(p-p_2)^2(p-p_3)^2} dp, \tag{33}$$

where $\mathbf{T}$ has been previously defined in (29), p_α are Stroh's eigenvalues and correspond to the three complex roots of the sixth-order polynomial equation $|\mathbf{\Gamma}(p)|^2 = 0$ with positive imaginary part [66]. In Eq (33),

$$\Phi_{ijklmn}(p) := \frac{D_{ij}(p)\hat{\Gamma}_{kl}(p)\hat{\Gamma}_{mn}(p)}{(p-\bar{p}_1)^2(p-\bar{p}_2)^2(p-\bar{p}_3)^2} \tag{34}$$

has been introduced together with the definition of $D_{ij} := \hat{n}_i\hat{n}_j + (\hat{n}_i\hat{m}_j + \hat{m}_i\hat{n}_j)p + \hat{m}_i\hat{m}_j p^2$, $\hat{\Gamma}_{jk}$ being the adjoint of Γ_{jk}, defined as $\Gamma_{pj}\hat{\Gamma}_{jk} = |\mathbf{\Gamma}(p)|\delta_{pk}$, where δ_{pk} is the Kronecker delta.

In order to provide an explicit boundary element formulation, Cauchy's residue theory for multiple poles is applied to evaluate the integrals in (28) and (33), so no integration is performed. Explicit formulae are summarised in [58] taking into account possible repeated Stroh's eigenvalues.

The integral equation (26) can be written as follows:

$$c_{ij}(\mathbf{x}')u_j(\mathbf{x}') + \sum_{e=1}^{N_e}\left\{\oint_{\partial\Omega_e}\check{T}_{ij}(\mathbf{x}',\mathbf{x})u_j(\mathbf{x})dS(\mathbf{x})\right\}$$

$$= \sum_{e=1}^{N_e}\left\{\int_{\partial\Omega_e}\check{U}_{ij}(\mathbf{x}',\mathbf{x})t_j(\mathbf{x})dS(\mathbf{x})\right\}, \qquad (35)$$

where the boundary is discretised into N_e elements of surface $\partial\Omega_e$ and the integrals over the boundary $\partial\Omega$ are replaced by the sum of the integrals over the surface of each element.

The functions u_i and t_i are approximated over each element $\partial\Omega_e$ using linear shape functions, as a function of the nodal values of the displacements and tractions, respectively:

$$u_i = \sum_{j=1}^{4}N_j(\xi,\eta)d_i^j, \quad t_i = \sum_{j=1}^{4}N_j(\xi,\eta)p_i^j. \qquad (36)$$

After discretising the boundary, Eq. (35) can be written as

$$\mathbf{Hd} = \mathbf{Gp}, \qquad (37)$$

where $\mathbf{d}$ and $\mathbf{p}$ contain the values of all nodal displacements and tractions, respectively.

3.2. *Contact discrete equations*

Assuming that the interface discretisation on $\partial\Omega_c$ is performed such that node to node contact is considered, each node on $\partial\Omega_c^1$ forms a contact pair I with one almost coincident node on $\partial\Omega_c^2$.

Performing a boundary element approximation of both solids, Eq. (37) can be rearranged according to the boundary conditions as follows:

$$\mathbf{A}_x^1\mathbf{x}^1 + \mathbf{A}_p^1\,\tilde{\mathbf{C}}^1\mathbf{p}_c = \mathbf{F}^1, \qquad (38)$$

$$\mathbf{A}_x^2\mathbf{x}^2 - \mathbf{A}_p^2\,\tilde{\mathbf{C}}^2\mathbf{p}_c z = \mathbf{F}^2. \qquad (39)$$

In Eqs. (38) and (39), $(\mathbf{x}^l)^T = [(\mathbf{x}_e^l)^T\ (\mathbf{d}_c^l)^T]$ $(l = 1,2)$ is the nodal unknowns vector that collects the contact nodal displacements $(\mathbf{d}_c^l)$ and the external unknowns $(\mathbf{x}_e^l)$. $\mathbf{A}_x^l$ is constructed with the columns of matrices $\mathbf{H}^l$ and $\mathbf{G}^l$, and $\mathbf{A}_p^l$ with the columns of $\mathbf{G}^l$ belonging to the contact nodal unknowns. Finally, vector $\mathbf{p}_c$ contains the normal and tangential contact tractions of every contact pair I, which are related with the nodal boundary element nodal tractions vector through the matrix $\tilde{\mathbf{C}}^l$ (i.e. $\mathbf{p}_c^1 = \tilde{\mathbf{C}}^1 \mathbf{p}_c$ and $\mathbf{p}_c^2 = -\tilde{\mathbf{C}}^2 \mathbf{p}_c$), described in [50].

The discrete gap for every contact pair I is approximated as

$$\mathbf{g}_c = \mathbf{C}_{g_n}(\mathbf{g}_o + \boldsymbol{\omega}) + (\mathbf{C}^2)^T \mathbf{x}^2 - (\mathbf{C}^1)^T \mathbf{x}^1, \tag{40}$$

where $\mathbf{g}_c$ contains the normal and tangential gap of every contact pair I, and $\mathbf{g}_o$ and $\boldsymbol{\omega}$ contain the initial gap and the wear gap of every contact pair I, respectively. $\mathbf{C}_{g_n}$ is an assembling matrix defined in [50].

Finally, the contact restrictions (5) and (16) are applied to every contact pair I:

$$(\mathbf{p}_n)_I - \mathbb{P}_{\mathbb{R}_-}(\ (\mathbf{p}_n)_I + r_n(\mathbf{g}_n)_I\ ) = 0, \tag{41}$$

$$(\mathbf{p}_t)_I - \mathbb{P}_{\mathbb{E}_\rho}(\ (\mathbf{p}_t)_I - r_t \mathbb{M}^2(\mathbf{g}_t)_I\ ) = \mathbf{0}, \tag{42}$$

where $\mathbf{p}_n$ and $\mathbf{p}_t$ contain the normal and tangential contact tractions of every contact pair I, respectively.

3.3. *Solution algorithm*

The quasi-static wear problem (38)–(42) can be solved using an iterative Uzawa scheme proposed by Rodríguez-Tembleque *et al.* [50–52, 67, 68] and considered by Feng *et al.* [69]. To compute the variables at the instant or load step (k), $\mathbf{z}^{(k)} = (\mathbf{x}^1, \mathbf{x}^2, \mathbf{p}_c, \mathbf{g}_c, \boldsymbol{\omega})^{(k)}$, when the variables at the previous instant $\mathbf{z}^{(k-1)}$ are known:

(I) Initialise $\mathbf{z}^{(0)} = \mathbf{z}^{(k-1)}$ (i.e. $\mathbf{p}_c^{(0)} = \mathbf{p}_c^{(k-1)}$ and $\boldsymbol{\omega}^{(0)} = \boldsymbol{\omega}^{(k-1)}$) and iterate using $(n = 0, 1, 2, 3, \ldots)$ index.

(II) Solve:

$$\begin{bmatrix} \mathbf{A}_x^1 & \mathbf{0} \\ \mathbf{0} & \mathbf{A}_x^2 \end{bmatrix} \begin{bmatrix} \mathbf{x}^1 \\ \mathbf{x}^2 \end{bmatrix}^{(n+1)} = \begin{bmatrix} -\mathbf{A}_p^1 & \tilde{\mathbf{C}}^1 \\ \mathbf{A}_p^1 & \tilde{\mathbf{C}}^2 \end{bmatrix} \mathbf{p}_c^{(n)} + \begin{bmatrix} \mathbf{F}^1 \\ \mathbf{F}^2 \end{bmatrix}^{(k)}.$$

(III) Compute contact gap:

$$\mathbf{g}_c^{(n+1)} = \mathbf{C}_{g_n}(\mathbf{g}_o^{(k)} + \boldsymbol{\omega}^{(0)}) + (\mathbf{C}^2)^T \mathbf{x}^{2(n+1)} - (\mathbf{C}^1)^T \mathbf{x}^{1(n+1)}.$$

(IV) Update contact tractions for every pair I:

$$(\mathbf{p}_n^{(n+1)})_I = \mathbb{P}_{\mathbb{R}_-}(\ (\mathbf{p}_n^{(n)})_I + r_n(\mathbf{g}_n^{(n+1)})_I\ ),$$

$$(\mathbf{p}_t^{(n+1)})_I = \mathbb{P}_{\mathbb{E}_\rho}(\ (\mathbf{p}_t^{(n)})_I - r_t\ \mathbb{M}^2(\Delta\mathbf{g}_t^{(n+1)})_I\ ),$$

where $(\Delta\mathbf{g}_t^{(n+1)})_I = (\mathbf{g}_t^{(n+1)})_I - (\mathbf{g}_t^{(0)})_I$ and $\rho = |(\mathbf{p}_n^{(n+1)})_I|$.

(V) Update accumulated wear depth:

$$(\boldsymbol{\omega}^{(n+1)})_I = (\boldsymbol{\omega}^{(0)})_I + |(\mathbf{p}_n^{(n+1)})_I| \, \|(\Delta \mathbf{g}_t^{(n+1)})_I\|_i.$$

(VI) Compute the error:

$$\Psi(\mathbf{p}_c^{(n+1)}) = \|\mathbf{p}_c^{(n+1)} - \mathbf{p}_c^{(n)}\|.$$

If $\Psi(\mathbf{p}_c^{(n+1)}) \leq \varepsilon$, the solution for the instant (k) is reached: $\mathbf{z}^{(k)} = \mathbf{z}^{(n+1)}$. Otherwise, return to (II) evaluating: $\mathbf{p}_c^{(n)} = \mathbf{p}_c^{(n+1)}$.

Once the solution at instant (k) is reached, the solution for the next instant is achieved by setting: $\mathbf{z}^{(0)} = \mathbf{z}^{(k)}$ and returning to (I).

4. Numerical examples

The capabilities of the proposed contact constitutive laws are applied to study the influence of the sliding direction (θ) and micromechanics (i.e. fibre orientation (φ) and fibre volume fraction (V_f)) on contact and wear in a carbon fibre-reinforced composite film. First, a steel sphere of radius $R = 100 \times 10^{-3}$ m is indented on a carbon FRP domain, whose dimensions are $2L_1 \times 2L_1 \times t$ (see Fig. 5(a)), being $L_1 = 50 \times 10^{-3}$ m and $t = \{0.5, 1, 2, 2.5, 5, 10, 20, 40, 80, 160\} \times 10^{-3}$ m. Then the indenter is subjected to normal load and cyclic tangential load, as Fig. 5(b) presents. For that purpose, the block is discretised by linear quadrilateral boundary elements, using 16×16 elements on the $L_o \times L_o$ potential contact zone $(L_o = 5 \times 10^{-3} \, \text{m})$, as Fig. 6 shows.

4.1. *Thick and thin films indentation response*

The spherical indenter is subjected to a normal displacement $g_{o,x_3} = -8 \times 10^{-5}$ m and a tangential translational displacement of module: $g_{o,t} = 2 \times 10^{-5}$ m, which forms an angle θ with axis x_1 (see Fig. 5(a)). The friction coefficients for a carbon FRP [4] are: $\mu_L = 0.4$, $\mu_T = 0.5$ and $\mu_N = 0.55$. In this example, the influence of the ratio of the contact radius a and the thickness of the film (i.e. a/t), the fibre orientation, the sliding direction and the micromechanics of FRP (i.e. the fibre volume fraction) in the contact variables are studied.

Micromechanics allows to estimate the mechanical properties of composite materials from the known values of the fibre and the matrix. There are different micromechanical approaches. The simplest approach is the *rule of mixtures*, but it fails to represent some of the properties with reasonable

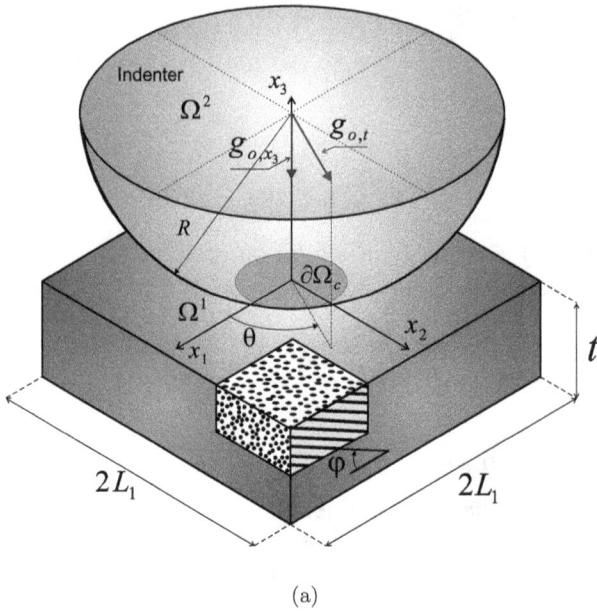

(a)

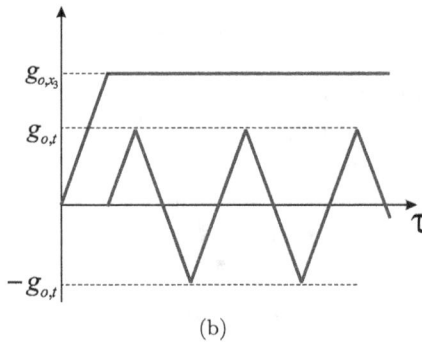

(b)

Fig. 5: (a) Spherical indenter over a fibre-reinforced composite film. (b) Normal load and cyclic tangential load. (These figures were reprinted from Rodríguez-Tembleque and Aliabadi [25] with permission from Elsevier.)

accuracy. A modified and more accurate micromechanical model was proposed by Hopkins *et al.* [41]. Also Halpin–Tsai proposed semi-empirical equations that have been applied for a long time to predict the properties of short-fibre composites. A detailed review of their derivation is given in Tucker *et al.* [70]. The carbon FRP considered is IM7 Carbon/8551–7, whose mechanical properties of fibre and matrix can be found in Kaddour *et al.* [71]. For the fibre, the longitudinal Young modulus is $E_{f1} = 276\,\text{GPa}$,

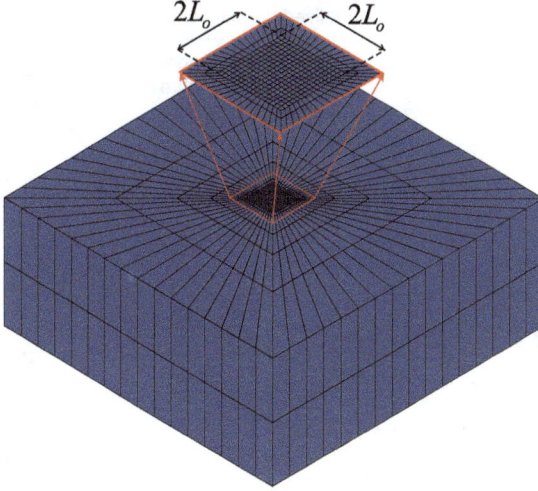

Fig. 6: Boundary element mesh details.

the transverse Young modulus are $E_{f2} = E_{f3} = 19\,\text{GPa}$, the in-plane shear modulus is $G_{f12} = 27\,\text{GPa}$, the transverse shear modulus is $G_{f23} = 7\,\text{GPa}$ and the Poisson ratios are $\nu_{f12} = \nu_{f13} = 0.2$. The mechanical properties of epoxy matrix are: $E_m = 4.08\,\text{GPa}$, $G_m = 1.478$ GPa and $\nu_m = 0.38$.

First, we are going to consider the fibre alignment $\varphi = 0°$ and the tangential load direction $\theta = 0°$. Figures 7(a) and (b) show the normal and tangential contact forces as functions of a/t and fibre volume fraction. Figure 7(a) shows normal contact forces (P) relative to the load for $V_f = 30\%$ (i.e. $P(V_f = 30\%)$). Figure 7(b) presents tangential contact forces (Q) relative to the load value at the transition to sliding regime for $V_f = 30\%$ (i.e. $\mu_L P(V_f = 30\%)$). It is seen from Fig. 7(a) that the transition occurs at around $a/t \simeq 1$. Moderate variations of the normal force occur on a thick film when $a/t \leq 0.1$, while high variations take place for $a/t > 1$. On the other hand, Fig. 7(b) shows that tangential contact force presents two transition points: $a/t \simeq 0.2$ and $a/t \simeq 1$. Those behaviours are observed both when the fibre volume fraction is of 45% and 60%.

The influence of the fibre orientation (φ) for $V_f = 45\%$ is studied in Fig. 8, where two thickness configurations are considered (i.e. $a/t = 0.0625$ and $a/t = 1.25$) and a higher tangential load is considered: $g_{o,t} = 6 \times 10^{-5}\,\text{m}$. For the normalised normal load, the largest loads occur for the normal fibre orientation ($\varphi = 90°$) and high differences can be observed for φ greater than 45°. The normalised tangential contact force presents a

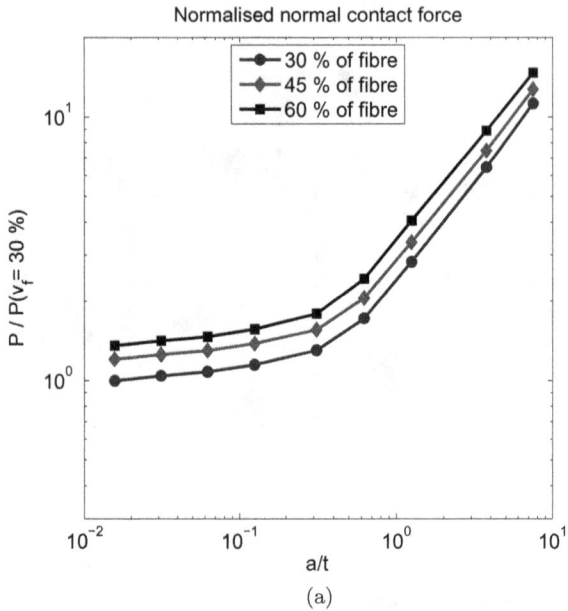

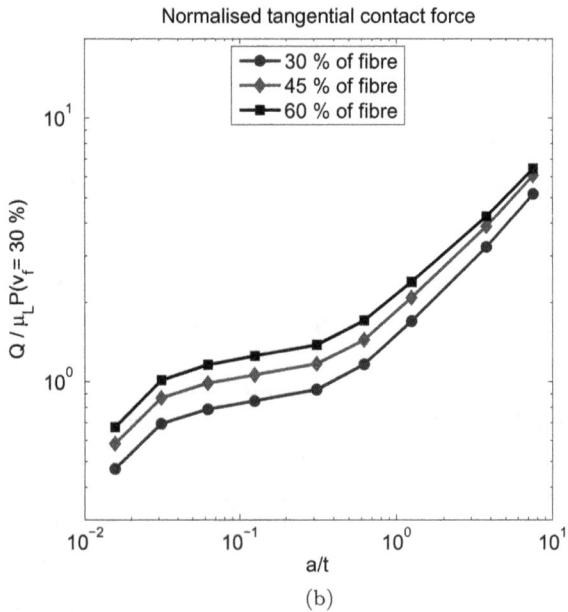

Fig. 7: Normalised indentation forces as functions of the contact radius and the block thickness: (a) Normal load and (b) tangential load, for different fibre volume fractions. (These figures were reprinted from Rodríguez-Tembleque and Aliabadi [25] with permission from Elsevier.)

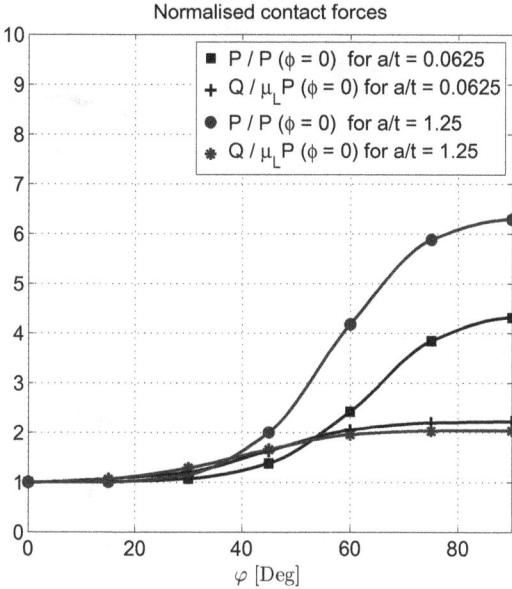

Fig. 8:   Normal and tangential contact forces as a function of the fibre orientation (φ). (This figure was reprinted from Rodríguez-Tembleque and Aliabadi [25] with permission from Elsevier.)

different behaviour. The largest discrepancies occur for a fibre orientation in the interval $[45°, 90°]$. The comparison between the forces for a thick film ($a/t = 0.0625$) and a thin film ($a/t = 1.25$) reveals that normal force increment is higher in thin films than in thick ones, but tangential forces are quite similar.

Finally, the behaviour of the proposed friction law is validated studying the influence of the sliding direction θ. Again, two thickness configurations are considered (i.e. $a/t = 0.0625$ and $a/t = 1.25$), the fibre volume fraction being 45%. Figure 9 presents the tangential load orientation $\theta = 0°$. Results reveal that the orientation of the fibres has an important effect on frictional response. For different sliding directions: $\theta = \{0°, 45°, 90°\}$, important discrepancies occur for a fibre orientation in the interval $[0°, 45°]$. However, the tangential force is not affected by θ for $\varphi = 90°$. This occurs because we recover the isotropic frictional behaviour ($\mu_1 = \mu_2 = \mu_N$) when $\varphi = 90°$.

4.2. *Thick and thin films under fretting wear conditions*

Previous indentation problem is now studied under fretting wear conditions. The indenter is subjected to a repeated cyclic tangential displacement of

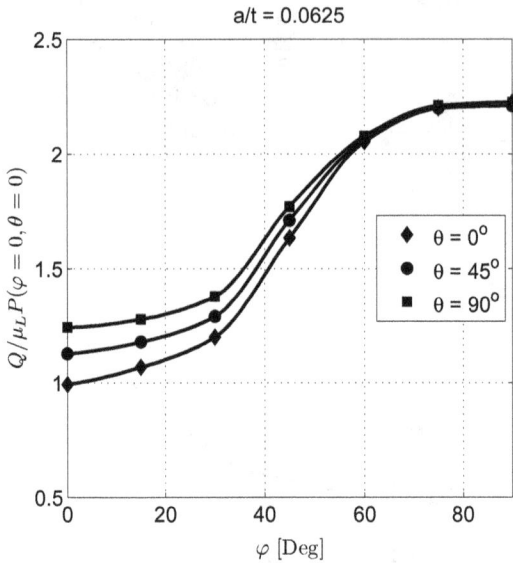

(a)

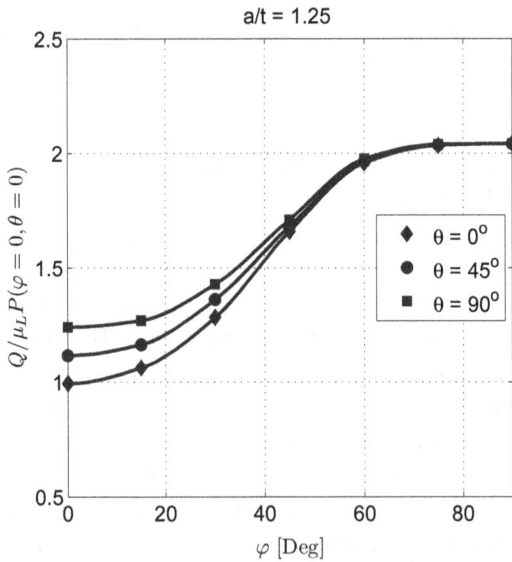

(b)

Fig. 9: Influence of fibre orientation and sliding direction in the tangential contact force for: (a) $a/t = 0.0625$ and (b) $a/t = 1.25$. (These figures were reprinted from Rodríguez-Tembleque and Aliabadi [25] with permission from Elsevier.)

module $g_{o,t}$ (see Fig. 5(b)) which forms an angle θ with axis x_1. Wear coefficients for the considered carbon FRP are $i_L = 5 \times 10^{-10}\,\mathrm{MPa}^{-1}$, $i_T = 6.25 \times 10^{-10}\,\mathrm{MPa}^{-1}$ and $i_N = 6.875 \times 10^{-10}\,\mathrm{MPa}^{-1}$. Two a/t ratios ($a/t = 0.0625$ and $a/t = 1.25$) are studied under partial slip and gross slip conditions. In partial slip, relative slip between the contacting bodies occurs on the brim of the contact region (a.k.a. slip region) and the contact points of each body move together without relative movement inside the brim (a.k.a stick region). Partial slip is considered for an isotropic friction law when $Q < \mu P$, where Q and P are the tangential and the normal contact forces, respectively, and μ, the static friction coefficient. When $Q = \mu P$, gross slip is formed, therefore, the whole contact region slips. Wear occurs in the slip region. For $g_{o,t} = 2 \times 10^{-5}\,\mathrm{m}$, we observe a sticky area inside the contact zone when $\varphi = 0$. However, all the contact points are in slip condition when $g_{o,t} = 6 \times 10^{-5}\,\mathrm{m}$. So both tangential load amplitudes will be considered to study wear.

Figures 10(a) and 10(b) present wear volume evolutions for partial slip conditions and different fibre volume fractions: $V_\mathrm{f} = \{30\%, 45\%, 60\%\}$, with the tangential load orientation $\theta = 0°$. Figure 10(a) shows wear volume evolution in a thick film ($a/t = 0.0625$) and Fig. 10(b) presents the evolution in a thin film ($a/t = 1.25$). In both cases, the fibre volume fraction has a moderate influence in the resulting wear volume (RWV). However, if we observe the resulting wear depth distribution in Figs. 10(c) and 10(d), the thin film presents higher wear depth values. The numerical aspects of the fretting wear computing scheme were presented and discussed in detail in the authors' previous works [50–52].

Wear volume evolutions under gross slip conditions are presented in Figs. 11(a) and 11(b). Both figures show the enormous influence of the fibre volume fraction in the RWV. This influence is even more significant in thin films (see Fig. 11(b)). This figure exhibits an increment of 33% in the RWV as a consequence of the fibre volume fraction increment. Figures 11(c) and 11(d) show the resulting wear depth distribution. Contrary to what was observed in the partial slip case, the thick film presents higher wear depth values and the contact zones are completely worn in both cases, because of the gross slip conditions. Under partial slip conditions, a small annular slip region was worn (Figs. 10(c) and 10(d)).

Figures 12–14 show the influence of the fibre orientation (φ) and the sliding direction (θ) on the resulting wear volume. For this numerical example, we have considered a fibre volume fraction: $V_\mathrm{f} = 45\%$, and a tangential

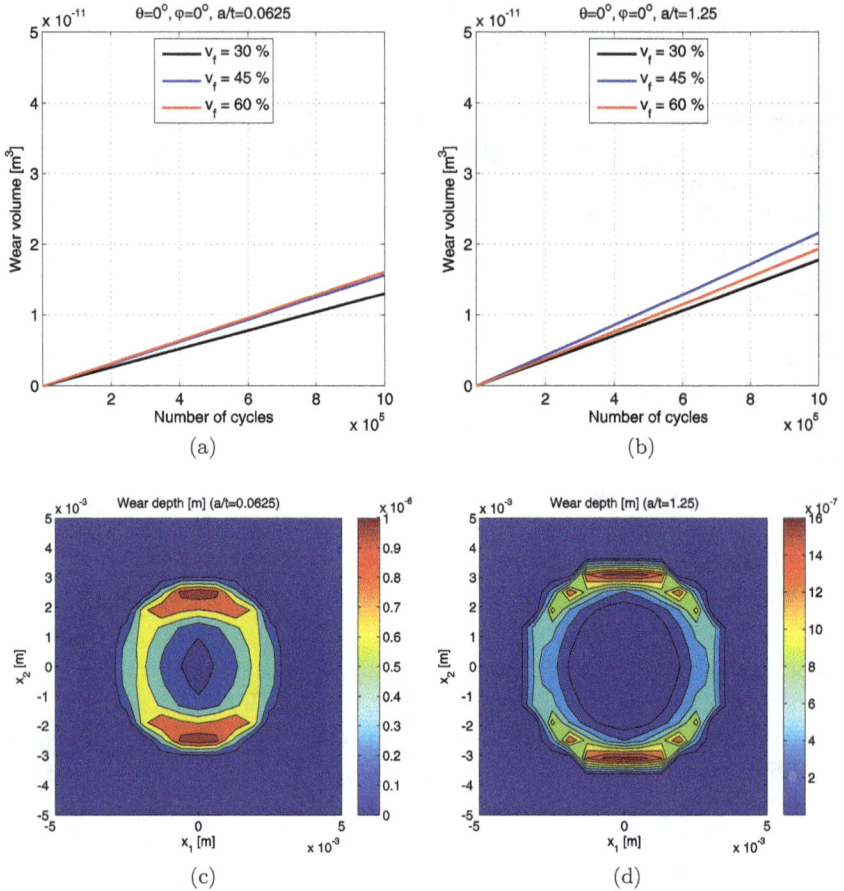

Fig. 10:   Influence of fibre volume fraction in wear volume evolution for: (a) $a/t = 0.0625$ and (b) $a/t = 1.25$, under partial slip conditions. Resulting wear depth distribution for: (c) $a/t = 0.0625$ and (d) $a/t = 1.25$. (These figures were reprinted from Rodríguez-Tembleque and Aliabadi [25] with permission from Elsevier.)

slip amplitude: $g_{o,t} = 6 \times 10^{-5}\,\mathrm{m}$. The resulting wear depth when the sliding direction $\theta = 0°$ is presented in Fig. 12, whereas results for the sliding angles $\theta = 45°$ and $\theta = 90°$ are presented in Figs. 13 and 14, respectively.

Examining these figures, it is found that the variation of the orientation of the sliding direction does not significantly affect the resulting wear depth distribution. Nevertheless, fibre's direction has an important effect on the magnitude of wear and on the wear topology. When fibres orientation is

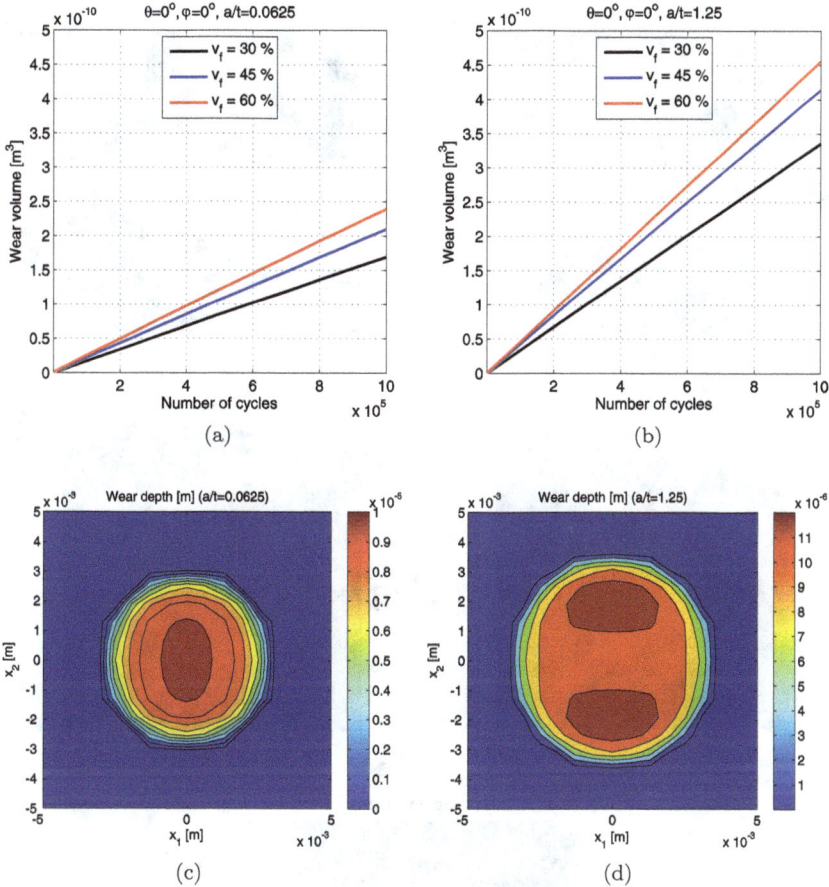

Fig. 11: Influence of fibre volume fraction in wear volume evolution for: (a) $a/t = 0.0625$ and (b) $a/t = 1.25$, under sliding wear conditions. Resulting wear depth distribution for: (c) $a/t = 0.0625$ and (d) $a/t = 1.25$. (These figures were reprinted from Rodríguez-Tembleque and Aliabadi [25] with permission from Elsevier.)

$\varphi = 0°$, contact zone is completely worn in all cases. However, if we increase the value of φ ($\varphi \geq 45°$), a smaller annular slip region is worn.

Finally, figures in Fig. 15 show the resulting normal contact pressure distribution for different values of $\varphi = \{0°, 45°, 90°\}$. They reveal that fibre's direction also has a significant effect on the magnitude of the contact tractions and the normal contact pressure distribution.

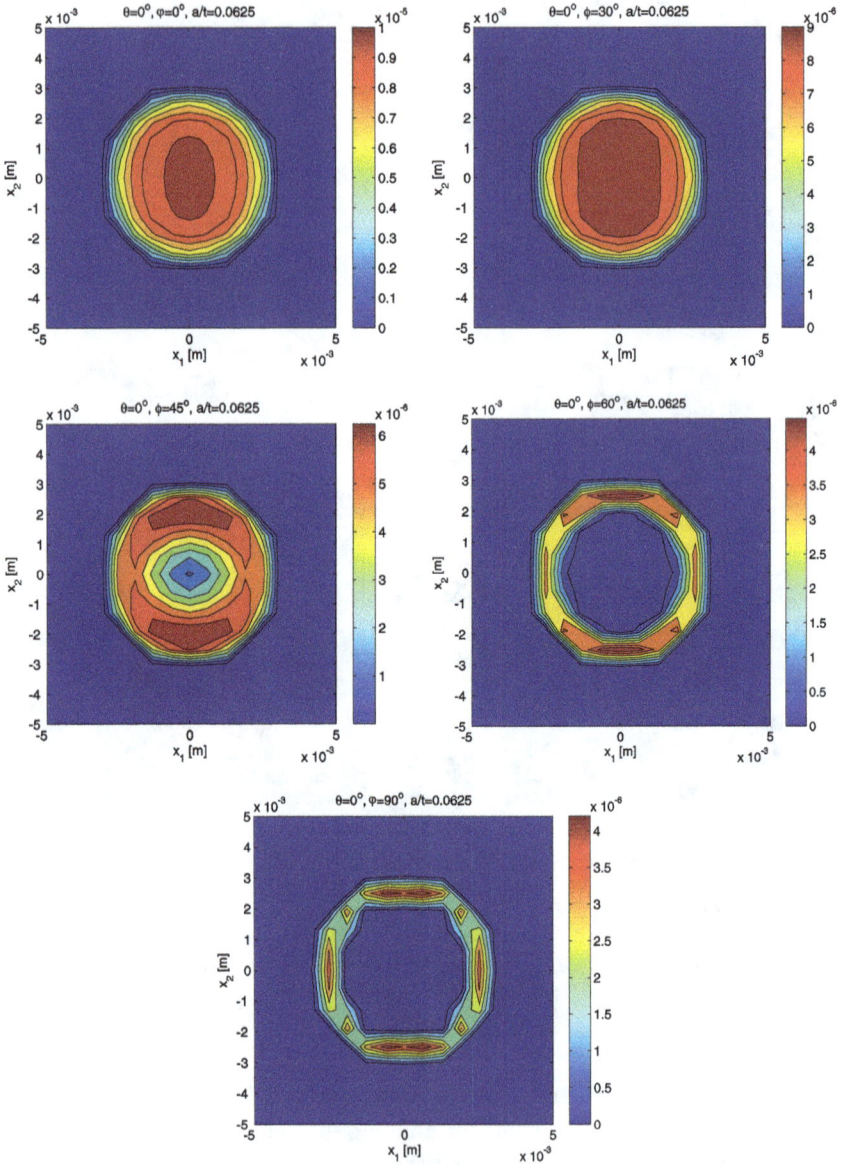

Fig. 12:   Influence of the fibre orientation in the resulting wear depth distribution for $\theta = 0°$. (These figures were reprinted from Rodríguez-Tembleque and Aliabadi [25] with permission from Elsevier.)

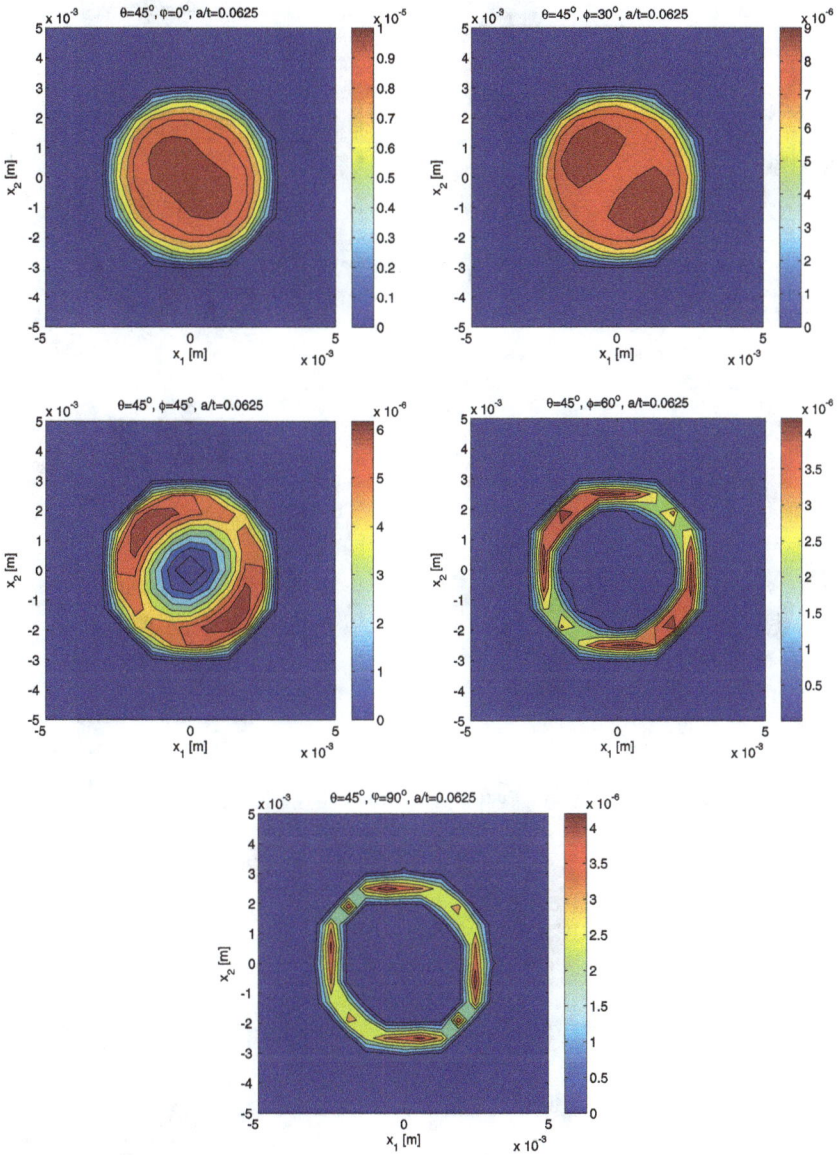

Fig. 13: Influence of the fibre orientation in the resulting wear depth distribution for $\theta = 45°$. (These figures were reprinted from Rodríguez-Tembleque and Aliabadi [25] with permission from Elsevier.)

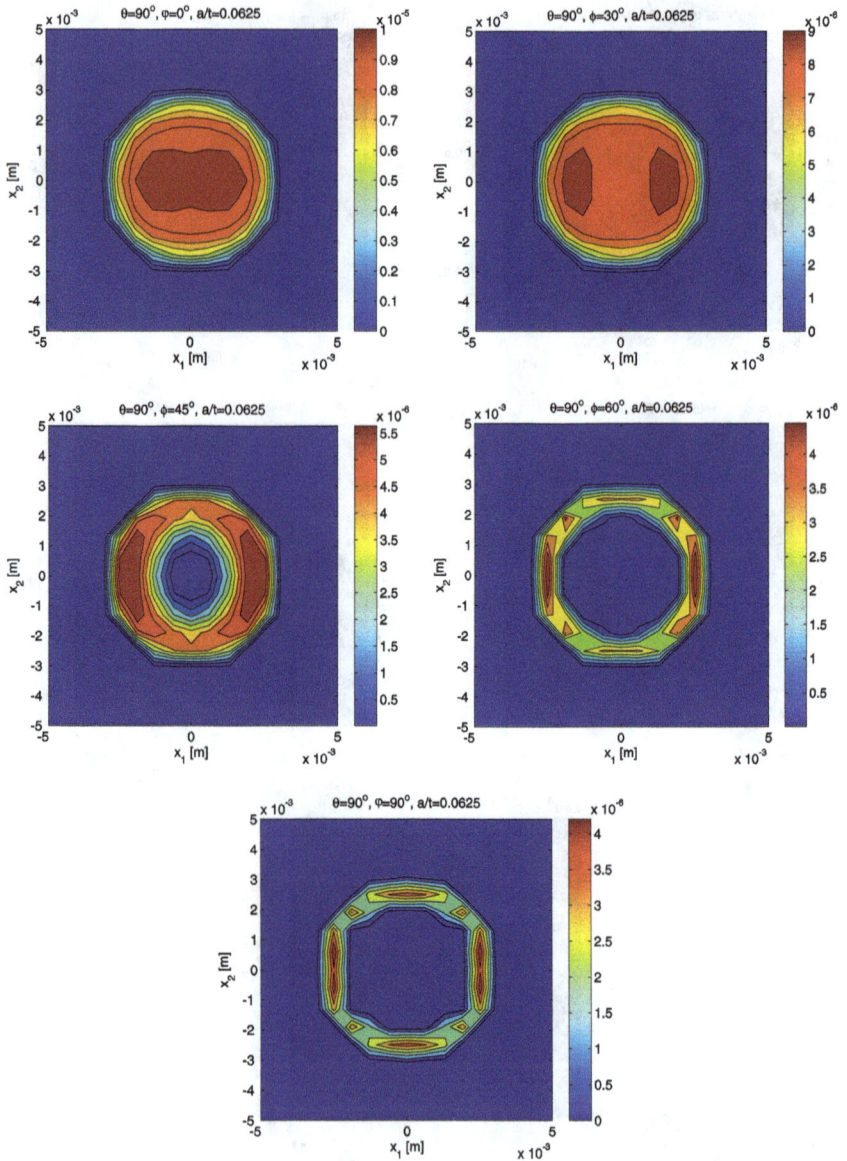

Fig. 14:   Influence of the fibre orientation in the resulting wear depth distribution for $\theta = 90°$. (These figures were reprinted from Rodríguez-Tembleque and Aliabadi [25] with permission from Elsevier.)

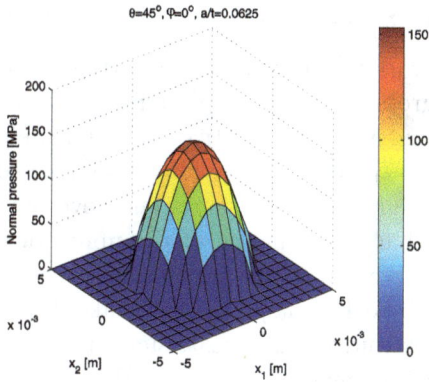

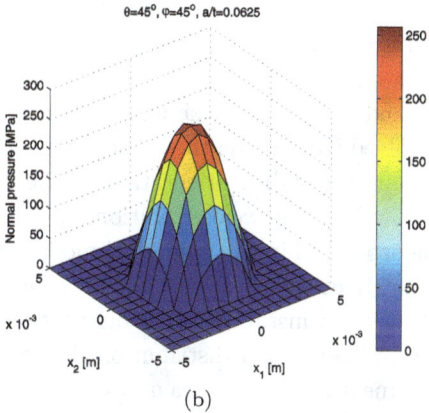

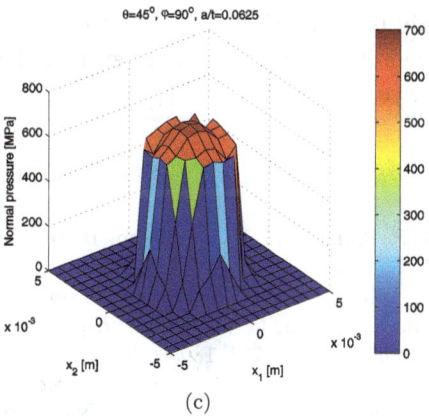

Fig. 15: Influence of the fibre orientation in the resulting normal contact pressure distribution for $\theta = 45°$. (These figures were reprinted from Rodríguez-Tembleque and Aliabadi [25] with permission from Elsevier.)

5. Concluding Remarks

A numerical framework for wear modelling fretting wear in fibre-reinforced composites under gross and partial slip conditions has been presented. The proposed friction and wear constitutive laws and their numerical implementation have been applied to study fibre-reinforced composite materials under different fretting wear conditions. The novelty with respect to the previous works is that the influence of the fibre orientation (φ) can be taken into account to compute the tribological properties on the surfaces (i.e. friction and wear coefficients). This fact makes it possible to have more realistic contact constitutive equations to simulate fibre-reinforced composite materials under frictional and wear conditions.

The methodology is based on the boundary elements, which prove to be very robust and suitable for this kind of tribological simulations. They obtain a good accuracy with a low number of elements, what makes the fretting wear resolution faster.

Several contact and fretting wear simulations on a carbon FRP block and film have been solved considering the influence of: the fibre orientation, the micromechanics and the sliding orientation on the normal and tangential contact force, as well as the wear evolution. Results reveal that these factors should be incorporated in the numerical models when FRP are subjected to fretting wear. In other cases, we could over- or underestimate wear and contact magnitudes. For instance, fibre's orientation has a fundamental effect on wear magnitude and wear distribution. For a constant tangential translational displacement $g_{o,t}$, fibre's orientation causes a completely worn contact zone or an annular worn region (see Fig. 12).

Finally, it should be noted that the presented methodology is useful in developing experimental indentation tests to extract the tribological properties of fibre-reinforced composites.

References

[1] M. S. Scholz, J. P. Blanchfield, L. D. Bloom, B. H. Coburn, M. Elkington, J. D. Fuller, M. E. Gilbert, S. A. Muflahi, M. F. Pernice, S. I. Rae, J. A. Trevarthen, S. C. White, P. M. Weaver, and I. P. Bond, The use of composite materials in modern orthopaedic medicine and prosthetic devices: a review, *Compos Sci Technol.* **71**(16), 1791–1803 (2011).

[2] L. C. Bank, *Composites for Construction — Structural Design with FRP Materials.* John Wiley & Sons, Inc., Hoboken, New Jersey, 2006.

[3] K. Friedrich, Polymer composites for tribological applications, *Adv Ind Eng Polym Res.* **1**, 3–39 (2018).

[4] N. Ohmae, K. Kobayashi, and T. Tsukizoe, Characteristics of fretting of carbon fibre reinforced plastics, *Wear.* **29**(3), 345–353 (1974).

[5] N. H. Sung and N. P. Suh, Effect of fiber orientation on friction and wear of fiber reinforced polymeric composites, *Wear.* **53**, 129–141 (1979).

[6] T. Tsukizoe and N. Ohmae, Friction and wear of advanced composite materials, *Fibre Sci Technol.* **18**, 265–286 (1983).

[7] M. Cirino, K. Friedrich, and R. B. Pipes, The effect of fiber orientation on the abrasive wear behavior of polymer composite materials, *Wear.* **121**, 127–141 (1988).

[8] O. Jacobs, K. Friedrich, G. Marom, K. Schulte, and H. D. Wagner, Fretting wear performance of glass-, carbon-, and aramid-fibre/epoxy and peek composites, *Wear.* **135**, 207–216 (1990).

[9] B. Vishwanath, A. P. Verma, and V. S. K. Rao, Effect of reinforcement on friction and wear of fabric reinforced polymer composites, *Wear.* **167**, 93–99 (1993).

[10] K. Friedrich, *Advances in Composite Tribology, Composite Materials Series, Vol. 8*, Elsevier, Amsterdam, 1993.

[11] G. Xian and Z. Zhang, Sliding wear of polyetherimide matrix composites i. influence of short carbon fibre reinforcement, *Wear.* **258**, 776–782 (2005).

[12] T. Larsen, T. L. Andersen, B. Thorning, A. Horsewell, and M. Vigild, Fretting wear performance of glass-, carbon-, and aramid-fibre/ epoxy and peek composites, *Wear.* **262**, 1013–1020 (2007).

[13] M. Sharma, I. M. Rao, and J. Bijwe, Influence of orientation of long fibers in carbon fiber-polyetherimide composites on mechanical and tribological properties, *Wear.* **267**, 839–845 (2009).

[14] J. JyotiKalita and K. K. Singh, Tribological properties of different synthetic fiber reinforced polymer matrix composites: a review, *IOP Conf Ser: Mater Sci Eng.* **455**, 012134 (2018).

[15] R. Vinayagamoorthy, Friction and wear characteristics of fibrereinforced plastic composites, *J Thermoplast Compos. Mater.* **33**, 828–850 (2020).

[16] X. Ning and M. R. Lovell, On the sliding friction characteristics of unidirectional continuous FRP composites, *J Tribol.* **124**, 5–13 (2002).

[17] X. Ning, M. R. Lovell, and C. Morrow, Anisotropic strength approach for wear analysis of unidirectional continuous FRP composites, *J Tribol.* **126**, 65–70 (2004).

[18] X. Ning, M. R. Lovell, and W.S. Slaughter, Asymptotic solutions for axisymmetric contact of a thin, transversely isotropic elastic layer, *Wear.* **260**, 693–698 (2006).

[19] R. C. Batra and W. Jaing, Analytical solution of the contact problem of a rigid indenter and an anisotropic linear elastic layer, *Int J Solids Struct.* **45**, 5814–5830 (2008).

[20] R. C. Batra and W. Jaing, Indentation of a laminated composite plate with an interlayer rectangular void, *Compos Sci Technol.* **70**, 1023–1030 (2010).

[21] J. Leroux and D. Nélias, Stick-slip analysis of a circular point contact between a rigid sphere and a flat unidirectional composite with cylindrical fibers, *Int J Solids Struct.* **148**, 3510–3520 (2011).

[22] C. Bagault and D. Nélias, Contact analyses for anisotropic half space: effect of the anisotropy on the pressure distribution and contact area, *J Tribol.* **134**, 1–8 (2012).

[23] C. Bagault, D. Nélias, M. C. Baietto, and T. C. Ovaert, Contact analyses for anisotropic half-space coated with an anisotropic layer: effect of the anisotropy on the pressure distribution and contact area, *Int J Solids Struct.* **50**, 743–754 (2013).

[24] J. Leroux, D. Nélias, and J.-A. Ruiz-Sabariego, Modélisation d'un contact frottant pour matériaux composites, *Matériaux Techniq.* **101**, 205 (2013).

[25] L. Rodríguez-Tembleque and M. Aliabadi, Numerical simulation of fretting wear in fiber-reinforced composite materials, *Eng Fract Mech* **168**, 13–27 (2016).

[26] J. Xiaoyu, Frictional contact analysis of composite materials, *Compos Sci Technol.* **54**, 341–348 (1995).

[27] M. Lovell, Analysis of contact between transversely isotropic coated surfaces: development of stress and displacement relationships using FEM, *Wear.* **194**, 60–70 (1998).

[28] K. Váradi, Z. Néder, J. Flöck, and K. Friedrich, Numerical and experimental contact analysis of a steel ball indented into a fibre reinforced polymer composite material, *J Mater Sci.* **33**, 841–851 (1998).

[29] K. Váradi, Z. Néder, J. Flöck, and K. Friedrich, Finite-element analysis of a polymer composite subjected to a ball indentation, *Compos. Sci. Technol.* **59**, 271–281 (1999).

[30] T. Goda, K. Váradi, B. Wetzel, and K. Friedrich, Finite element simulation of the fiber–matrix debonding in polymer composites produced by a sliding indentor: Part II — normally oriented fibers, *J Compos Mater.* **38**, 1583–1606 (2004).

[31] T. Goda, K. Váradi, B. Wetzel, and K. Friedrich, Finite element simulation of the fiber–matrix debonding in polymer composites produced by a sliding indentor: Part II — parallel and anti-parallel fiber orientation, *J Compos Mater.* **38**, 1607–1618 (2004).

[32] M. H. Aliabadi, *The Boundary Element Method Vol2: Applications in Solids and Structures.* John Wiley & Sons, Chichester, 2002.

[33] J. Varna, F. París, and J. C. del Cano, The effect of crack-face contact on fiber/matrix debonding in transverse tensile loading, *Compos Sci Technol.* **57**, 523–532 (1997).

[34] E. Graciani, V. Mantic, F. París, and A. Blázquez, Weak formulation of axi-symmetric frictionless contact problems with boundary elements: Application to interface cracks, *Comput Struct.* **83**, 836–855 (2005).

[35] E. Graciani, V. Mantic, F. París, and J. Varna, Numerical analysis of debond propagation in the single fibre fragmentation test, *Compos Sci Technol.* **69**, 2514–2520 (2009).

[36] L. Távara, V. Mantic, E. Graciani, and F. París, BEM analysis of crack onset and propagation along fiber-matrix interface under transverse tension using a linear elastic-brittle interface model., *Eng Anal Bound Elem.* **35**, 207–222 (2011).

[37] V. Mallardo and C. Alessandri, Inverse problems in the presence of inclusions and unilateral constraints: a boundary element approach, *Comput Mech.* **26**, 571–581 (2000).

[38] L. Rodríguez-Tembleque, F. C. Buroni, R. Abascal, and A. Sáez, Analysis of FRP composites under frictional contact conditions, *Int J Solids Struct.* **50**, 3947–3959 (2013).

[39] L. Rodríguez-Tembleque, F. C. Buroni, and A. Sáez, Numerical study of polymer composites in contact, *CMES.* **96**, 131–158 (2013).

[40] L. Rodríguez-Tembleque and M. Aliabadiz, Friction and wear modelling in fiber-reinforced composites, *CMES.* **102**, 183–210 (2014).

[41] D. Hopkins and C. Chamis, A unique set of micromechanics equations for high-temperature metal matrix composites, in *Testing Technology of Metal Matrix Composites*, ed. N. Adsit and P. DiGiovanni (West Conshohocken, PA: ASTM International, 1988), 159–175. https://doi.org/10.1520/STP25950S

[42] A. Curnier, A theory of friction, *Int J Solids Struct.* **7**, 637–647 (1984).

[43] R. Michalowski and Z. Mróz, Associated and non-associated sliding rules in contact friction problems, *Int J Solids Struct.* **30**, 259–276 (1978).

[44] Z. Mróz and S. Stupkiewicz, An anisotropic friction and wear model, *Int J Solids Struct.* **31**, 1113–1131 (1994).

[45] Z. Mróz, S. Kucharski, and I. Páczelt, Anisotropic friction and wear rules with account for contact state evolution, *Wear.* **296-397**, 1–11 (2018).

[46] Z.-Q. Feng, M. Hjiaj, G. de Saxcé, and Z. Mróz, Effect of frictional anisotropy on the quasistatic motion of a deformable solid sliding an a planar surface, *Comp Mech.* **37**, 349–361 (2006).

[47] Z.-Q. Feng, M. Hjiaj, G. de Saxcé, and Z. Mróz, Influence of frictional anisotropy on contacting surfaces during loading/unloading cycles, *Inter J Eng Tribol.* **41**, 936–948 (2006).

[48] L. Rodríguez-Tembleque and R. Abascal, Fast FE-BEM algorithms for orthotropic frictional contact, *Int J Numer Methods Eng.* **94**, 687–707 (2013).

[49] I. R. McColl, J. Ding, and S. B. Leen, Finite element simulation and experimental validation of fretting wear, *Wear.* **256**, 1114–1127 (2004).

[50] L. Rodríguez-Tembleque, R. Abascal, and M. H. Aliabadi, A boundary element formulation for wear modeling on 3D contact and rolling contact problems, *Int J Solids Struct.* **47**, 2600–2612 (2010).

[51] L. Rodríguez-Tembleque, R. Abascal, and M. H. Aliabadi, A boundary element formulation for 3D fretting-wear problems, *Eng Anal Boundary Elem.* **35**, 935–943 (2011).

[52] L. Rodríguez-Tembleque, R. Abascal, and M.H. Aliabadi, Anisotropic wear framework for 3D contact and rolling problems, *Comput Meth Appl Mech Eng.* **241**, 1–19 (2012).

[53] I. Paczelt, S. Kucharski, and Z. Mróz, The experimental and numerical analysis of quasi-steady wear processes for a sliding spherical indenter, *Wear.* **274-275**, 127–148 (2012).

[54] I. Paczelt and Z. Mróz, Solution of wear problems for monotonic and periodic sliding with p-version of the finite element method, *Comput Meth Appl Mech Eng.* **249-252**, 75–103 (2012).

[55] S. Stupkiewicz, An ALE formulation for implicit time integration of quasi-steady-state wear problems, *Comput Meth Appl Mech Eng.* **260**, 130–142 (2013).

[56] J. F. Cavalieri and A. Cardona, Three-dimensional numerical solution for wear prediction using a mortar contact algorithm, *Int J Numer Methods Eng.* **96**, 467–486 (2013).

[57] J. Lengiewicz and S. Stupkiewicz, Efficient model of evolution of wear in quasi-steady-state sliding contacts, *Wear.* **303**, 611–621 (2013).

[58] F. C. Buroni, J. E. Ortiz, and A. Sáez, Multiple pole residue approach for 3D BEM analysis of mathematical degenerate and non-degenerate materials, *Int J Numer Methods Eng.* **86**, 1125–1143 (2011).

[59] F. Tonon, E. Pan, and B. Amadei, Green's functions and boundary element method formulation for 3D anisotropic media, *Comput Struct.* **79**, 469–482 (2001).

[60] C. Y. Wang and M. Denda, 3D BEM for general anisotropic elasticity, *Int J Solids Struct.* **44**, 7073–7091 (2007).

[61] E. Pan and B. Amadei, A 3-D boundary element formulation of anisotropic elasticity with gravity, *Appl Math Modell.* **20**, 114–120 (1996).

[62] A. Sáez, M. P. Ariza, and J. Dominguez, Three-dimensional fracture analysis in transversely isotropic solids, *Eng Anal Bound Elem.* **20**, 287–298 (1997).

[63] M. Loloi, Boundary integral equation solution of three-dimensional elasto-static problems in transversely isotropic solids using closed-form displacement fundamental solutions, *Int J Numer Meth Eng.* **48**, 823–842 (2000).

[64] L. Távara, J. E. Ortiz, V. Mantic, and F. Paris, Unique real-variable expressions of displacement and traction fundamental solutions covering all transversely isotropic elastic materials for 3D BEM, *Int J Numer Meth Eng.* **74**, 776–798 (2008).

[65] V. G. Lee, Explicit expressions of derivatives of elastic green's functions for general anisotropic materials, *Mech Res Comm.* **30**, 241–249 (2003).

[66] T. C. T. Ting, *Anisotropic Elasticity.* Oxford University Press, Oxford, 1996.

[67] L. Rodríguez-Tembleque, F. Buroni, and A. Sáez, 3D BEM for orthotropic frictional contact of piezoelectric bodies, *Comput Mech.* **56**, 491–502 (2015).

[68] L. Rodríguez-Tembleque, F. Buroni, A. Sáez, and M. Aliabadi, 3D coupled multifield magneto-electro-elastic contact modelling, *Int J Mech Sci.* **114**, 35–51 (2016).

[69] P. Ning, Z. Feng, J. Rojas, Y. Zhou, and L. Peng, Uzawa algorithm to solve elastic and elastic–plastic fretting wear problems within the bipotential framework, *Comput Mech.* **62**, 1327–1341 (2018).

[70] C. L. T. III and E. Liang, Stiffness predictions for unidirectional short-fiber composites: Review and evaluation, *Compos Sci Technol.* **59**, 655–671 (1999).

[71] A. S. Kaddour and M. J. Hinton, Input data for test cases used in benchmarking triaxial failure theories of composites, *J Compos Mater.* **54**, 2295–2312 (2012).

Index

Computational and Experimental Methods in Structures

(Continued from page ii)

www.ingramcontent.com/pod-product-compliance
Lightning Source LLC
Chambersburg PA
CBHW050554190326
41458CB00007B/2032